Genetics

Genetics

D. J. COVE

Fellow of Trinity Hall, Cambridge,
University Lecturer in Genetics

CAMBRIDGE
At the University Press 1971

Published by the Syndics of the Cambridge University Press

Bentley House, 200 Euston Road, London NW1 2DB
American Branch: 32 East 57th Street, New York, N.Y.10022

© Cambridge University Press 1971

Library of Congress Catalogue Card Number: 75–160089

ISBNs:
      0 521 08255 2 clothbound
      0 521 09663 4 paperback

Printed in Great Britain by
William Clowes & Sons Limited
London, Colchester and Beccles

Contents

Preface

It may be thought that in choosing the title 'Genetics' for this book, I have been arrogant, but I have been persuaded that an ostensibly more modest title such as 'An Introduction to Genetics' demands a much lengthier text than this, perhaps even in several volumes. This book is not of course a comprehensive textbook of Genetics. It is an elementary and introductory text which omits completely several important fields including for example the genetics of populations. It is based on the lectures which I give, as part of the Biology of Cells course, to first year undergraduates reading Natural Sciences at Cambridge. Many of these students become acquainted with genetics for the first time through these lectures, and I have found that the somewhat unconventional approach to Mendelism by way of the genetics of haploid organisms has proved particularly satisfactory for such students.

There are many whom I should thank for the help given to me in writing this book, but in particular I should like to thank Professor J. M. Thoday for the help he gave me in devising the original lecture course, and so indirectly in the writing of this book. I should also like to thank Dr K. J. R. Edwards, Mr M. Isaacs and Dr D. W. Mac-Donald for their helpful criticism of the manuscript, Dr S. A. Henderson for providing the photograph of meiosis, Dr. M. Ashburner for providing the photograph of chromosome puffing, Mrs J. Merry and Miss B. Short for typing the manuscript.

D. J. COVE

Trinity Hall, Cambridge
June 1971

1. Patterns of inheritance I: haploid organisms

To understand the mechanism of inheritance, it is first necessary to consider the ways in which cells and organisms reproduce. Two different processes of reproduction are found at the cellular level, and for many species at the organism level too. By far the most common process of cell reproduction found in eukaryotic* and prokaryotic* organisms is the process whereby a cell results from the division of a parent cell, and in turn divides to give rise to daughter cells. There is very good evidence that the information inherited by these daughter cells is almost always identical to that possessed by the mother cell. Much of the evidence for this will be found in later chapters of this book. This process is usually called **asexual reproduction**, and it is the process of asexual cell reproduction which accounts for the growth of bacterial cultures and also for the development of a human from a single cell to the final adult organism with a cell content of about 10^{13}–10^{14} cells.

In eukaryotic organisms, a cell can sometimes be formed, not by asexual reproduction, but instead by the fusion of two special cells. This process involving cell fusion is the central one of the second process of reproduction, **sexual reproduction**. Prokaryotic organisms also show sexual reproduction, but the mechanisms involved are more diverse and more complex. These will be described in this book but only after the mechanism of inheritance in eukaryotes has been dealt with. Returning to sexual reproduction in eukaryotic organisms, the specialised cells which fuse are called **gametes**, and the cell which is formed by their fusion is called a **zygote**.

Confusion often arises when these two processes of reproduction are considered at the cellular and organism level. Where the whole organism consists of a single cell then these two levels are the same

---

* These terms are defined briefly in the *Index of Definitions and Glossary* on page 207.

and no confusion can occur. It is with multicellular organisms, that confusion can occur. At the organism level humans only reproduce sexually; two parents are involved, each of which supplies one gamete to form a zygote which develops into another individual. However, from a cellular point of view by far the most common method of cell reproduction in humans is the asexual one. In some multicellular organisms asexual reproduction occurs not only at the cellular level, but also at the organism level. Some examples of this will probably be familiar. The strawberry plant, *Fragaria grandiflora*, for example sends out long shoots at the ends of which a new plant can develop, which will become independent of the parent plant when the shoots wither. No gametes are involved in this process, nor therefore is a zygote, only asexual cell reproduction has occurred and yet two individuals have arisen where before there was only one.

The fungus, *Aspergillus nidulans*, is another example of a eukaryotic organism which can reproduce both sexually and asexually. As it is relatively simple, it will be used in this book to illustrate some of the basic facts about the mechanism of inheritance and so it is first necessary to consider its life cycle. *Aspergillus* is a fairly common mould often found growing on food, and is related to *Penicillium*, the mould used to make penicillin, and also to the moulds used in the making of certain cheeses. *Aspergillus* consists of long tubular cells, which are typically eukaryotic in structure. Each cell is however very large and contains many nuclei. It is possible that during the course of evolution *Aspergillus* lost the cell walls that divided each cell with its single nucleus from its neighbours, and so the multinucleate cells found today were formed. These cells are called **hyphae**, and branch as they grow. Collectively the branching cells are called a **mycelium**. The mycelium grows by the process of asexual cell reproduction, but *Aspergillus* can also reproduce asexually at the organism level. The cells of the mycelium produce, also by asexual cell reproduction, enormous numbers of spores, called **conidiospores** (or sometimes **conidia**). It is these spores which give *Aspergillus* and other moulds their characteristic green colour. Each square centimetre of *Aspergillus* mycelium produces about 2×10^7 spores, and so a mycelium the size of a petri-dish (i.e. 9 cm in diameter) produces about 10^9 spores. Any one of these spores can germinate and grow to a similar size producing another 10^9 spores, in about five days. The mycelium which develops from a conidiospore is almost always identical to that from which the spore came. This then is the asexual form of reproduction shown by *Aspergillus*.

In addition to this asexual method of reproduction, *Aspergillus* also reproduces sexually, producing gametes which fuse to form zygotes, which in turn undergo cell division to give rise to another form of spore, the **ascospore**. Ascospores germinate to give rise to an *Aspergillus* mycelium. The life cycle of *Aspergillus* is represented diagrammatically in figure 1. Although there is more to the life cycle than this, and one of the complications will be dealt with later in this book, this is a sufficient account to enable us to consider the behaviour of *Aspergillus* in simple breeding experiments.

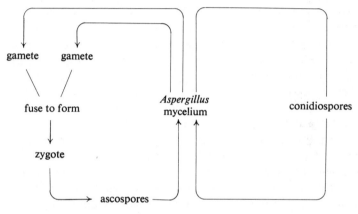

Figure 1. Simplified life cycle of *Aspergillus nidulans*.

Having outlined the life cycle of *Aspergillus*, it is now possible to consider in what way breeding experiments can be used to gain an insight into the mechanism of heredity. But the first thing that must be noted is that breeding experiments can only be useful provided we have recognisably different strains. If two identical strains of *Aspergillus* are crossed together, the vast majority of the progeny which develop from the ascospores are identical to one another, and to the parents. Clearly this tells us nothing about the mechanism of inheritance that we did not know already. However, variant strains can be obtained. These arise spontaneously, but only at a very low frequency. It was stated earlier that conidiospores, the spores involved in the asexual form of reproduction of *Aspergillus*, almost always produce a mycelium identical to the one upon which they were borne. In fact among these conidiospores can be found a very small proportion which give rise to recognisably different types of mycelium, which produce conidiospores almost all of which give rise to the new type of mycelium. The differences are therefore inherited, and it appears there must have been a change in the genetic information. Such altered strains are said to be **mutant**, and the alteration in the genetic information which has occurred is called a **mutation**. Mutation can occur at any time and so ascospores too may very rarely be mutant. The rate at which mutations occur in *Aspergillus* varies depending on the type of mutation involved, but a rate of 1 in 10^7 is fairly typical. Thus 1 in 10^7 conidiospores from a normal green-spored (**wild type**) *Aspergillus* mycelium, may give rise to a mycelium which bears yellow rather than green conidiospores. However the mutation rate can be greatly increased by various treatments. These include ultraviolet irradiation, X-rays, γ-rays and certain chemicals. Such treatments are said to be **mutagenic**, and their nature, and the nature of the mutations they produce tell us a great deal about the process of mutation. Chapter 5 (page 64) of this book deals with the mutation process in greater detail.

The effect of a mutation on the cell and organism is also very informative and by its study we now know much about how the genetic information is used. Although this too will be considered in more detail in later chapters of this book, it will be helpful for an understanding of the mechanism of inheritance to consider briefly at this point the main effects of mutation. One type of mutation, that which leads to the production of yellow spores has been mentioned already. Other mutations in *Aspergillus* can cause fawn, chartreuse, pale green or white conidiospores to be produced. It is also possible to obtain mutations which may have different effects on morphology. We can for example obtain strains with a slower growth rate, which in a given time, give rise to a mycelium of a diameter smaller than that of a wild-type strain. Another type of mutation may render the strain resistant to various chemicals normally toxic to the wild type. The most frequently used mutations in *Aspergillus* are however those called **biochemical** or sometimes **nutritional mutations**.

The first extensive study of biochemical mutation was by G. W. Beadle and E. L. Tatum in the period around 1940. These workers did not use *Aspergillus* for their studies but another fungus *Neurospora crassa*. Both these fungi are able to grow on a relatively simple medium consisting of inorganic salts and a sugar. Beadle and Tatum found that it was possible, by irradiating conidiospores with u.v. or X-rays so as to increase the mutation rate, to obtain strains which were only able to grow provided a biochemical was added to the medium. Many different types of mutant were found. Some for example would only grow when a particular amino acid was added to the medium, some only when a vitamin was added, and so on. They reasoned that the mutation must have led to the strain carrying it being unable to synthesise the particular biochemical that was required for growth. A more detailed analysis, testing the mutant strains for growth on the known precursors of the required bio-chemicals, lead Beadle and Tatum to propose that biochemical mutations resulted in the absence of a single enzyme from the cells of the strain which carried them. Suppose a biochemical X is syn-thesised from a precursor A, through intermediates B, C and D, a four-step process catalysed by four enzymes, a, b, c and d as follows:

$$A \xrightarrow{a} B \xrightarrow{b} C \xrightarrow{c} D \xrightarrow{d} X$$

A typical mutation might lead to a strain being able to grow provided X **or** D **or** C were added to the medium but not if B or A were added. This growth pattern is consistent with the strain lacking enzyme b. Another mutation might lead to a strain carrying it being able to grow only if X were added, and not if either A, or B, or C or D. This strain appears to lack enzyme d. These examples are illustra-tive of many actual cases and often the conclusion can be confirmed by direct enzyme measurements. As a result of their work Beadle and Tatum proposed that genetic information was used by cells to synthesise enzymes. They were not in fact the first workers to have proposed this, but the value of their contribution lies in that they showed that the genetic determination of enzymes was a much more widespread phenomenon than it had been thought to be hitherto.

We are now equipped with sufficient detail to consider what can be deduced about the process of inheritance from simple breeding experiments.

One of the simplest breeding experiments which can be performed with *Aspergillus* is to cross a mutant strain to the wild type. The results of such a cross are simple but nevertheless informative. If for example a yellow-spored mutant strain is crossed to the green-spored wild-type strain, the resulting ascospores develop into progeny of two types only, resembling one or other of the two parents. Furthermore these two types are produced in approximately equal numbers. Similar results are obtained if other types of mutant are used. When a strain which requires the vitamin biotin in order to grow is crossed to the wild type, equal numbers of biotin requirers and non-requirers occur among the progeny. Breeding programmes of this sort can be represented formally as has been done in figure 2.

Straightway something has been learnt about the process of information transfer. No blending of the wild-type and mutant charac-

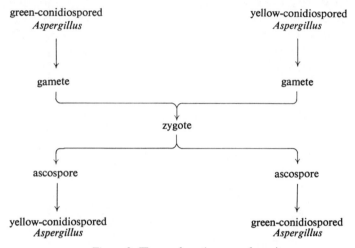

Figure 2. (For explanation, see above.)

ters has occurred. The progeny do not have yellowy-green conidiospores, but only either the original parental yellow or green colours. The progeny are said to show **segregation** of the two parental types. Note also that another possibility has not arisen, that is that no progeny occur which are capable of producing both yellow and green conidiospores. From this experiment it can be reasoned that:

(1) Each strain of *Aspergillus* contains with respect to the difference between the production of yellow and green conidiospores, only one type of information, that is it can either contain information leading to the production of yellow spores or to the production of green spores.

(2) Since the zygotes give rise to equal numbers of ascospores of the two parental types, it is likely that the parents contributed equally to the zygote, that is the contribution of the two gametes to the zygote is equal. The zygote must therefore contain two types of genetic information with respect to the difference between yellow spores and green spores.

(3) Since the progeny do not produce yellow-green conidiospores these two alternative types of genetic information contained in the zygote cannot get mixed in any way.

(4) Since the ascospores produced from the zygote give rise to mycelium bearing only yellow *or* green conidiospores, the information in the zygote must be sorted out precisely during the formation of the ascospores, so that each ascospore receives only one of the two alternative types of information.

These conclusions can be represented symbolically. Suppose we call the two alternative pieces of genetic information *a* and *A*. The breeding scheme can now be represented as in figure 3.

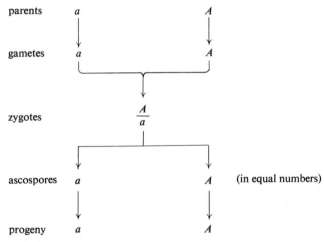

Figure 3. (For explanation, see above.)

Cells which have only one copy of a particular type of genetic information are said to be **haploid**. Cells having two copies are called **diploid**. In the scheme in figure 3 therefore, only the zygote is diploid. Notice also that the scheme proposes that there are two different types of cell division. The more common type, corresponding to asexual cell reproduction, results in the daughter cell receiving the same information content as that of the mother cell from which it came. This is the process by which a haploid cell gives rise to haploid cells. However during the process whereby the diploid zygote gives rise to haploid ascospores an orderly reduction of information content must occur. This type of cell division whereby a diploid cell can give rise to haploid cells is called a **reduction division**.

At this point, it is convenient to introduce the term **gene**. It is best to define a gene as a unit of hereditary information. It will be seen later that there is little point in being more explicit than this. No more precise definition is possible as it would not cover everything which is usually called a gene. In the breeding scheme represented in figure 3, we can say that A and a represent alternative forms of the same gene. Such alternative forms are called **alleles**.

Now let us examine what occurs when we do a slightly more complex breeding experiment. What happens if two different mutant strains are crossed together? The results are again simple. If for example a yellow-conidiospored strain of *Aspergillus* is crossed with a slow-growing strain, four classes of progeny are obtained in equal numbers. These classes are the two parental types, one producing yellow conidiospores but having the normal growth rate, and the other producing wild-type, green conidiospores but having a slow growth rate. Two new types are also obtained, one with green conidiospores and a normal growth rate, identical in fact to the wild-type strain from which the two parental strains were derived by mutation, and the other with yellow conidiospores and a slow growth rate, which can be described as being doubly mutant. These two new non-parental types are called **recombinants** (see figure 4).

The first thing to note from this experiment is that with respect to the individual characters involved, i.e. rate of growth and spore colour, all the conclusions that were drawn from the previous breeding experiments still hold. There is for example no blending of information with respect to an individual character, and no yellowy-green conidiospored progeny, or ones having an intermediate rate of growth occur. It can again be said that the individual alleles of the two genes involved remain distinct from one another during the process of zygote formation and reduction division.

What in addition emerges from this experiment is that the formation of a zygote by the gametes of the two different strains, followed by the production of ascospores must provide an opportunity for the shuffling or **reassortment** of genetic information affecting different characters. Progeny are obtained which have some information derived from one parent and some from the other. A complete set of genetic information is called a **genome**, and so this last conclusion can be rephrased in more technical terms by saying that progeny derive their genes from both of their parents' genomes. The single character cross did not enable any predictions to be made as to whether this would be so. It might have been that not only the information affecting a single character but also the entire parental genomes remained separated in the zygote with no opportunity for mixing to occur.

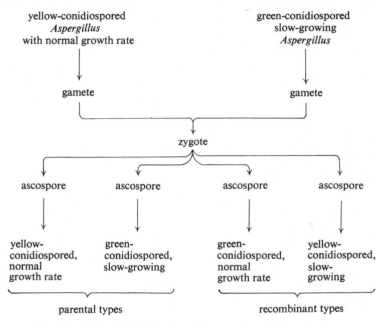

Figure 4. (For explanation, see opposite.)

Since all four classes of progeny resulting from a two character cross occur at equal frequencies, it can be argued that in the re-assortment of genetic information which takes place during the reduction division by which the zygote gives rise to ascospores, information for each particular character is distributed independently of information for other characters. In other words, it can be said that genes for different characters assort independently. This point becomes clearer when the cross is represented symbolically (see figure 5). It will be seen later however that not all pairs of genes segregate independently of one another (see Chapter 3, Linkage and the chromosome theory).

From now on, a more informative type of gene symbol will be used. Unfortunately various different forms of genetic symbolism are in use for different organisms. The most common form, which will be used for most of the examples described in this book, represents the wild-type allele of a particular gene by a +. The mutant allele is represented by a letter or letters often being an abbreviation of the mutant character. Thus the mutant allele leading to yellow

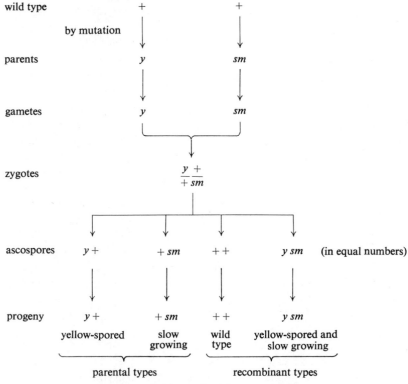

Figure 5. (For explanation, see p. 9.)

conidiospores is called *y*, and the mutant allele causing slow growth, and hence small mycelia, *sm*. If a strain is designated + it is usual to assume that it is wild type with respect to all the genes in its genome. Similarly if a strain carries a single mutation, and is designated for example *y*, it is assumed that it is wild type with respect to the rest of the genes in its genome. Occasionally when it is wished to draw attention to the fact that a strain carries a wild-type allele of a particular gene, this is done by adding the mutant symbol as a superscript to the +. Thus $+^y$ indicates that a strain carries a wild-type allele of the gene in which mutation can occur to give the y allele. These new symbols are used in figure 5, which represents the cross between the yellow-conidiospored strain and the slow-growing strain symbolically.

If the scheme represented in figure 5 is correct, it should be possible to take a yellow-spored slow-growing recombinant progeny strain and cross it either to the original wild type, or to one of the wild-type progeny and to obtain exactly the same classes of progeny in the same proportions as in the original cross. This cross would be $y\ sm \times +$. These are indeed the results obtained when such crosses are carried out.

Aspergillus was not used in the original breeding experiments, which led to the conclusions about the nature of the mechanism of inheritance which have been drawn in this chapter. These original experiments, which were carried out by a monk, Gregor Mendel, working in Brno, in Czechoslovakia, used the garden pea, *Pisum sativum*. Mendel published his classic papers on inheritance in 1866, although his work attracted little attention until the beginning of this century. In them he proposed all the principles already mentioned in this chapter; particularly that no blending occurred for individual character differences, and that information for different characters could be distributed independently to the progeny. This was a very remarkable achievement, especially as the results which Mendel obtained with his breeding experiments using a flowering plant were quite different from those which are obtained with *Aspergillus*. These results will be described in detail in the next chapter, which will also show how they can be reconciled with the results of breeding experiments with fungi.

2. Patterns of inheritance II: diploid organisms

Mendel used for his experiments, strains of pea which were commercially available. These differed in a number of respects and he was able to observe the results of crossing strains differing by one or more characters. However he first checked that each strain was genetically pure, by self-pollinating it, i.e. crossing within a strain, and showing that its progeny were identical to itself. Such strains are said to be **true breeding**. He studied the inheritance of a number of character differences, but since he obtained similar results for each, it is necessary to describe the results of only one such experiment. When Mendel crossed a true-breeding strain whose flowers were purple, with a true-breeding white-flowered strain, he found that all the progeny from the cross had purple flowers. To understand how such a result can be reconciled to the breeding behaviour of *Aspergillus* which has already been described it is necessary to take the experiment with the pea further. This Mendel did, he self-pollinated the purple-flowered progeny from the first cross, and found that they gave rise to progeny three quarters of which had purple flowers and one quarter of which had white. These experiments are summarised in figure 6. Mendel's actual results in the second generation were 705 purple-flowered plants, and 224 white-flowered plants, which is a ratio of 3.1 to 1.

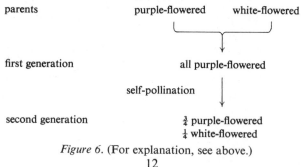

parents	purple-flowered	white-flowered
first generation	all purple-flowered	
self-pollination		
second generation	¾ purple-flowered	
	¼ white-flowered	

Figure 6. (For explanation, see above.)

12

The first conclusion which can be drawn is that the hybrid progeny from the first cross (these are generally called the F_1 **generation**) although apparently resembling the purple-flowered parent strain cannot be identical to it, for whereas the parent strain upon self-pollination gives rise only to progeny like itself, this is not true of the hybrids. These, although they have purple flowers, must also contain information for white flowers in some latent form, because they regularly give rise to white-flowered progeny. The situation is clarified if the second generation progeny (usually called the F_2 **generation**) are investigated further. Mendel self pollinated the members of the F_2 generation and found that whereas the quarter which produced white flowers were true breeding just like the white-flowered parent strain, the purple-flowered progeny were of two types. One third of the purple-flowered F_2 progeny (i.e. $\frac{1}{4}$ of the total progeny) were true breeding, but the remaining two-thirds (i.e. $\frac{1}{2}$ of the total progeny) behaved just like the purple-flowered F_1 generation, giving rise to purple-flowered and white-flowered progeny in the ratio of three to one.

The simplest way to explain these results is to postulate that each pea plant contains two pieces of information for any one character, i.e. is diploid. In a true-breeding strain these two pieces of information are similar. During gamete formation, the information content is halved, so that the gametes are haploid. Gamete production must then involve a reduction division. When gametes from two true-breeding strains fuse to form a zygote this will be diploid again, and will contain an equal amount of information from each parent. The same will be true of the hybrid which develops from this zygote. It must next be proposed that although this hybrid contains information both for purple flowers and white flowers it produces purple flowers. When this hybrid comes to produce gametes by reduction division, two types will be produced in equal numbers. It is easier to see the result of random fusion of these gametes among themselves, if symbols are again adopted. With a cultivated species like the pea, it is a little difficult to know what is wild type, but the purple-flowered allele will rather arbitrarily be designated $+$, and the white-flowered w. The two parental strains are diploid and may therefore be represented $\frac{+}{+}$ for the purple-flowered strain, and $\frac{w}{w}$ for the white-flowered strain. The breeding scheme can then be represented as in figure 7.

It is convenient at this juncture to introduce a few more technical terms. A diploid cell in which the two copies of a particular gene are represented by the same allele is said to be **homozygous** for that allele. If however the two alleles are different, it is said to be **heterozygous**.

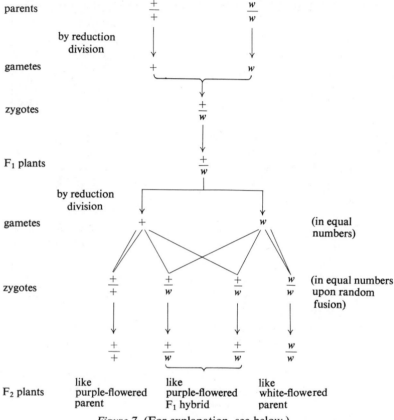

Figure 7. (For explanation, see below.)

A homozygous strain will upon self-pollination breed true, whereas a heterozygous strain will show segregation among its progeny. If a heterozygous strain, $\frac{A}{a}$, resembles the strain homozygous for the A allele, i.e. $\frac{A}{A}$, then the A allele is said to be **dominant** to the a allele, or conversely the a allele is said to be **recessive** to the A allele. In the example just described and represented symbolically in figure 7, the $+$ allele is dominant to the w allele.

The hypothesis which Mendel put forward to explain the results from his single character crosses, which have been summarised in figure 7, can be tested. If the scheme is correct, then the cross between the purple-flowered F_1 hybrid $\left(\text{heterozygous } \dfrac{+}{w}\right)$ and the true-breeding white-flowered parental strain $\left(\text{homozygous } \dfrac{w}{w}\right)$ should give equal numbers of $\dfrac{+}{w}$ (purple-flowered) and $\dfrac{w}{w}$ (white-flowered) progeny (see

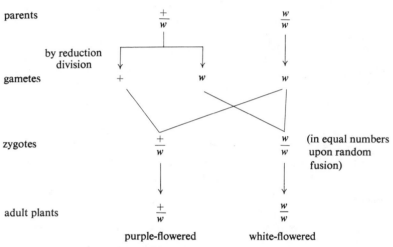

Figure 8. (For explanation, see above.)

figure 8 for breeding scheme). Mendel found that this was so. He obtained 84 purple-flowered and 81 white-flowered progeny, which is in close agreement to the predicted 1 : 1 ratio. This type of cross, where an F_1 hybrid is crossed to one of the parent strains is called a **back cross**. The special case where the F_1 hybrid is crossed to a strain homozygous for the recessive alleles involved (this need not necessarily be a parent strain) is called a **test cross**.

The theory of dominance which proposes that the heterozygous strain resembles one of the two homozygous strains which gave rise to it, may at first sight seem to be proposing a rather unlikely situation. However it will be recalled that mutation commonly leads to the absence of an enzyme. It is probable that the w allele has arisen from the $+$ allele by mutation, and that strains homozygous for the w allele are unable to make an enzyme necessary for the synthesis of purple flower pigment. The heterozygote, $\frac{+}{w}$ will contain one wild-type

allele, and so will be able to make some functional enzyme. It will therefore be able to synthesise purple pigment. The possession of one wild-type allele by a heterozygote does not necessarily ensure that the heterozygote will resemble the homozygous wild-type. There are other examples, some also involving flower pigment, where the heterozygote is intermediate between the two parental strains. Thus when a true-breeding red-flowered strain of *Antirrhinum* is crossed to a true-breeding white-flowered strain, the F_1 hybrids are pink-flowered. The F_2 from this F_1 consists of red-flowered, pink-flowered and white-flowered plants in the ratio 1 : 2 : 1, and so it is possible here to observe directly the segregation which Mendel could only detect fully by further breeding experiments. In this case presumably the heterozygote with its single wild-type allele cannot make enough functional enzyme to catalyse the synthesis of the full amount of flower pigment, with the result that its flowers are pink not red.

The breeding behaviour of *Aspergillus* indicated that its cells are haploid, with the exception of the diploid zygote which undergoes reduction division to restore the normal haploid state. In the pea we have deduced that its cells are diploid, with the exception of the gametes, which must therefore be produced by reduction division, and which fuse to form a zygote, so restoring the normal diploid state. These two apparently contrasting situations can be resolved by representing their life cycles as has been done in figure 9.

In *Aspergillus*, the diploid phase is eliminated, leaving the zygote as the only diploid cell, whereas in the pea the haploid phase is much reduced, so that nearly all cells are diploid. The life cycle of all eukaryotic organisms conforms basically to this scheme. Man and most other animals have no haploid phase, the gametes being the only haploid cells. Most plants show to some extent both a diploid and a haploid phase, and in some cases, as for example is found in the ferns,

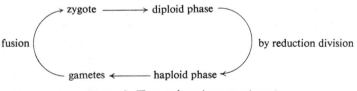

Figure 9. (For explanation, see above.)

these two phases are independent of one another, and morphologically quite distinct. The familiar fern plant is in fact the diploid phase, giving rise to spores by reduction division. These germinate to develop into small insignificant haploid plants, which produce gametes. Organisms, such as the pea or man, whose cells are almost all diploid, are commonly called **diploid organisms**, whereas organisms such as *Aspergillus* are described as **haploid organisms**.

The outcome of a cross involving two different characters in a diploid organism has not yet been considered. In *Aspergillus* it will be remembered that the genes for the two characters were distributed independently of one another to the progeny (see figure 5, page 10). For the situation in a diploid organism, Mendel's data for the pea can again be used. He crossed a true-breeding strain having round and yellow seed with a true-breeding strain having wrinkled and green seed. All the F_1 hybrid seeds were round and yellow. It can therefore be deduced that roundness is dominant to wrinkledness, and yellowness to greenness. When the F_1 hybrids developing from this seed were self-pollinated, F_2 seed was obtained as follows:

315 round, yellow seeds
101 wrinkled, yellow seeds
108 round, green seeds
32 wrinkled, green seeds.

Firstly, it can be seen that each character has segregated with approximately a 3 : 1 ratio, as flower colour did in the first pea cross described. The segregation for seed shape is 423 round to 133 wrinkled, and for seed colour 416 yellow to 140 green. Secondly, it can be seen that the two characters, seed shape and seed colour have segregated independently of one another. Thus among yellow seeds there is a 3 : 1 segregation of round to wrinkled, similarly among wrinkled seeds there is a 3 : 1 segregation of yellow to green and so on. The result of superimposing two 3:1 ratios, is a 9:3:3:1 ratio [i.e. 3(3:1):1(3:1)]. It will be seen that the results which were obtained by Mendel are close to this ratio. It therefore appears that two pairs of genes are segregating independently of one another. This situation can be represented symbolically. If the round allele of the gene for seed shape is represented by +, and the wrinkled allele by *wr*, and if the yellow allele of the gene for seed colour is represented by +, and the green allele by *gr*, the two parental strains, which because they are

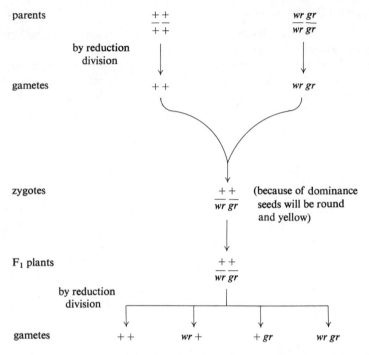

These can pair randomly to form sixteen types of zygote. The origin of these is given in the table below. The appearance of the F_2 seed which will develop from each zygote, taking dominance into account, is given below each zygote.

gametes from F_1	$+\,+$	$wr\,+$	$+\,gr$	$wr\,gr$
$+\,+$	$\dfrac{+\,+}{+\,+}$ round yellow	$\dfrac{wr\,+}{+\,+}$ round yellow	$\dfrac{+\,gr}{+\,+}$ round yellow	$\dfrac{wr\,gr}{+\,+}$ round yellow
$wr\,+$	$\dfrac{+\,+}{wr\,+}$ round yellow	$\dfrac{wr\,+}{wr\,+}$ wrinkled yellow	$\dfrac{+\,gr}{wr\,+}$ round yellow	$\dfrac{wr\,gr}{wr\,+}$ wrinkled yellow
$+\,gr$	$\dfrac{+\,+}{+\,gr}$ round yellow	$\dfrac{wr\,+}{+\,gr}$ round yellow	$\dfrac{+\,gr}{+\,gr}$ round green	$\dfrac{wr\,gr}{+\,gr}$ round green
$wr\,gr$	$\dfrac{+\,+}{wr\,gr}$ round yellow	$\dfrac{wr\,+}{wr\,gr}$ wrinkled yellow	$\dfrac{+\,gr}{wr\,gr}$ round green	$\dfrac{wr\,gr}{wr\,gr}$ wrinkled green

Figure 10. (For explanation, see opposite.)

true-breeding for both characters, must be homozygous for both genes, can be represented as

$$\frac{+}{+} \quad \text{or} \quad \frac{+^{wr} +^{gr}}{+^{wr} +^{gr}}$$

for the strain with round yellow seeds, and

$$\frac{wr \; gr}{wr \; gr}$$

for the strain with wrinkled green seeds. The breeding scheme is represented in full in figure 10.

The scheme represented in figure 10, predicts that the $\frac{9}{16}$ of the F_2 progeny with round yellow seeds will be genetically heterogeneous. Only $\frac{1}{9}$ would be homozygous for the round and yellow alleles and would therefore breed true. Mendel showed by further breeding experiments that these predictions were upheld. He also test crossed his F_1 hybrids to the strain which was homozygous for the recessive alleles of the genes involved. In this case, this is the parent with wrinkled green seed. He predicted that he would obtain the four possible classes (round yellow; wrinkled yellow; round green; and wrinkled green) in equal numbers. The test cross is represented symbolically in figure 11. In his cross, Mendel obtained 55, 49, 51 and

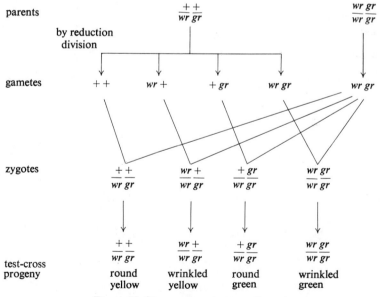

Figure 11. (For explanation, see above.)

52 respectively of the four classes, which was in close agreement to the predicted 1 : 1 : 1 : 1 ratio.

Mendel was an extremely thorough investigator, and he also studied the inheritance of three different characters in the same cross, where he obtained in the F_2 generation a 27 : 9 : 9 : 9 : 3 : 3 : 3 : 1 ratio which it will be seen is the expected result of superimposing a 3 : 1 ratio onto a 9 : 3 : 3 : 1 ratio [3(9 : 3 : 3 : 1) : 1(9 : 3 : 3 : 1)].

Mendel's results, which have been repeated since in a wide range of eukaryotic organisms both plant and animal, established that the genetic information behaved as though it were made up of discreet particles which were inherited independently of one another. It will be seen later that neither of these generalisations hold, but nevertheless Mendel's work was of the highest importance both because it established the basic facts about the mechanism of inheritance and because it served as a model for further investigations. Soon after it began to attract general attention at the beginning of this century, the first major exception to Mendel's results emerged. It was found that not all unrelated characters were inherited independently. The significance of this and what it led to, is discussed in the next chapter.

3. Linkage and the chromosome theory

The first reported example of two unrelated characters failing to segregate independently was published in 1906 by W. Bateson, E. R. Saunders and R. C. Punnett, who used the sweet pea, *Lathyrus odoratus*, for their experiments. They found that certain genes affecting flower colour and shape of pollen grain tended to be inherited together. This phenomenon is called **linkage**. Linkage is in fact a fairly common phenomenon, and Mendel was perhaps fortunate not to encounter it among the characters he studied. It is possible to quantify the degree of linkage shown by two genes. For the first example of how this is done, we will return to *Aspergillus*. It will be recalled that in the cross involving the two characters, conidiospore colour and growth rate, the four possible types of progeny were obtained at equal frequencies (see figure 5, page 10). At the same time, the point was made that not all pairs of genes segregated independently. If a yellow-conidiospored strain is crossed to a strain requiring the vitamin biotin, it is found that a considerable excess of parental types over recombinant types is obtained. This cross is represented symbolically in figure 12, where typical results for the segregation obtained are also given. It will be seen that each character behaves in the usual way, giving a 1 : 1 ratio of the two alternative types among the progeny (119 yellow-conidiospored : 113 green-conidiospored and 111 biotin-requirers: 121 non-requirers), but that the two characters have not assorted independently.

For a given pair of genes, the recombinant types always appear at about the same frequency. Furthermore it is not important which alleles are present in which parents. In the cross in the example, the yellow-conidiospore and the biotin-requirement alleles were carried by different parents, that is they are said to be in **repulsion**. If a yellow-conidiospored, biotin-requiring recombinant strain and a wild-type strain are crossed together, so that the alleles are distributed in the alternative manner, that is the yellow-conidiospore and the biotin-

21

requirement alleles are in **coupling**, the new parental types still predominate over the new recombinant types to about the same extent. Typical numbers of progeny from such a cross are; parental: wild-type, 142; yellow-conidiospored biotin-requiring, 130; recombinant: yellow conidiospored, 9; biotin-requiring 8. The degree

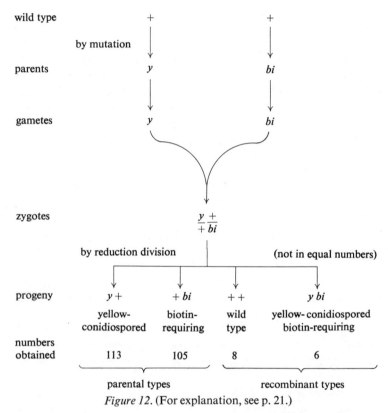

Figure 12. (For explanation, see p. 21.)

of linkage, the tendency of alleles from the same parent strain to be inherited together, can be quantified. The metric used is the **recombination fraction**. This is given by the simple expression:

$$\text{recombination fraction} = \frac{\text{number of progeny of recombinant types}}{\text{total number of progeny}}\%.$$

The data for the first cross, where the yellow-conidiospore and the biotin-requirement alleles were in repulsion, give a recombination fraction of 6.0% [(14 × 100)/232]. The second cross where these alleles are in coupling gives 5.9 % [(17 × 100)/289]. Other combinations of genes give different values, and all degrees of linkage are found, although the maximum value that can be obtained will be 50%, as when two genes assort independently in *Aspergillus* half the progeny will be recombinant, and half parental. Recombination fractions at the lower end of the scale pose another problem. Suppose in a cross involving two different characters, no recombinant progeny are obtained. It is impossible to tell whether this is because, even though two different genes were involved, no recombinants were detected as the size of the sample examined was too small, or whether this is because the two characters are different effects of the same gene. In the latter case, the two characters affected in a mutant strain, might be the result of the absence of a single enzyme. When a given gene affects more than one character it is said to have **pleiotropic** effects. It is difficult to be certain whether the absence of recombinants in a cross is the result of **pleiotropy**, although the better the understanding of the biochemical basis of the relevant characters, the better will be the position for deciding whether pleiotropy is involved. Mendel himself encountered strain differences which were almost certainly the effects of a gene showing pleiotropy. His strains with purple flowers always had brown seed coats and reddish spots on the leaf axils, whereas his strains with white flowers had colourless seed coats, and no reddish spots on the leaf axils. Mendel assumed that these character differences were all interconnected, and he observed that they were always inherited in the same combinations. It seems probable that the white-flowered strains lacked an enzyme involved in pigment bio-synthesis which lead to alterations in all three of the characters involved.

The measurement of recombination fractions in diploid organisms is not quite so straight forward as it is in haploid organisms. Linkage is apparent by an excess of the parental classes and a deficiency of the recombinant classes among the F_2 progeny. Characteristically, ratios of the >9 : <3 : <3 : >1 type are obtained. However it is easiest to measure recombination fractions if test crosses are used. For two character crosses where the genes segregate independently, the four possible classes of progeny occur with equal frequencies (see figure 11, page 19). If the genes involved show linkage however, the parental classes will be more frequent than the recombinant classes, and a recombination fraction can be calculated as in *Aspergillus*, by determining the proportion of recombinants among the

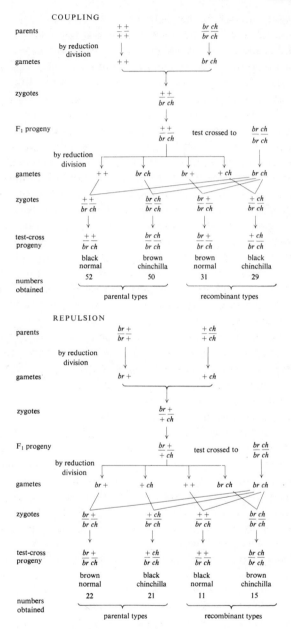

Figure 13. (For explanation, see opposite.)

progeny, and expressing them as a percentage. As an example, a rabbit cross will be used. If a rabbit from a true-breeding strain with black normal fur is crossed to a rabbit from a true-breeding brown chinchilla strain, all the F_1 progeny are black with normal fur. We can say, therefore, that brown is recessive to black, and that chinchilla is recessive to normal. When the F_1 progeny are test crossed to a doubly recessive strain (i.e. brown and chinchilla), the four possible classes are not obtained in equal frequencies, but the parental types are more common than the recombinant types, indicating that the two genes involved are linked. In this cross the brown and chinchilla alleles are in coupling, but similar linkage is obtained from the repulsion cross where the parents will be true breeding for brown normal fur and black chinchilla. The F_1 progeny will again be black with normal fur, and in the test cross the new parental types will predominate. These two crosses, and typical results obtained from them are shown in figure 13, where the alleles are represented as follows: black, $+$; brown, br; normal fur, $+$; and chinchilla, ch.

Again the first thing to note with these test crosses, is that each character difference has segregated in a 1 : 1 ratio (e.g. 81 black: 81 brown in coupling cross, 32 black : 37 brown in repulsion cross), confirming that a single gene only is involved for each character. The linkage values for the two crosses are in agreement. The coupling cross gives a recombination fraction of 37.0% $[(60 \times 100)/162]$, and for the repulsion cross 37.7% $[(26 \times 100)/69]$.

It is often found that the members of a pair of linked genes also show linkage to further genes, so that a linked series can be established. The recombination fractions between members of a linked series, which is called a **linkage group**, bear a special relationship to one another. The nature of this relationship will be illustrated with an example from the fruit fly, *Drosophila melanogaster*, which is by far the best genetically characterised animal there is. *Drosophila* has contributed enormously to our knowledge of genetics, and it was using *Drosophila* that T. H. Morgan in the period 1910-20 established the chromosome theory of linkage. The symbols for the recessive alleles involved in the example, together with their effect are as follows:

> cv absence of cross veins in wings,
> v vermilion eye colour,
> g garnet eye colour,

ct cut, an abnormal wing shape,
ec echinus, abnormal morphology of body bristles,
sc scute, absence of bristles from the scutelar region,
vg vestigial, wing size greatly reduced.

The various recombination fractions between the genes involved can be determined by the test cross procedure outlined above. Typical values for the recombination fractions obtained are as follows:

cv and *v*	22.8%
v and *g*	10.2%
cv and *g*	31.6%
v and *ct*	15.2%
cv and *ct*	7.9%
v and *ec*	36.2%
cv and *ec*	13.7%
v and *sc*	41.3%
cv and *sc*	21.3%
cv and *vg*	49.8%
v and *vg*	50.1%.

The last two values involving the *vg* gene are the maximum possible, indicating that *vg* is inherited independently of *cv* and *v*. It is probable that *vg* is not a member of the linkage group to which all the other genes belong. If the recombination fractions involving these other genes are examined, it will be seen that they bear an approximately linear relationship to one another. This can be illustrated using the first three values. The recombination fraction for *cv* and *g* (31.6%) is approximately equal to the sum of the values for *cv* and *v* (22.8%) and *v* and *g* (10.2%). Similarly, if the values for the *cv*, *v* and *ct* genes are considered it will be seen that the *cv* and *v* value (22.8%) is approximately equal to the sum of the values for *cv* and *ct* (7.9%) and *v* and *ct* (15.2%). In this way the genes can be placed in a linear order, and their relative positions 'mapped' using the recombination fractions. The complete map is shown below.

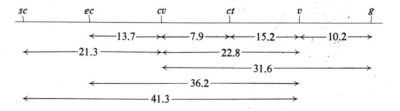

The values obtained for *vg* were consistent with this gene not being involved in the linkage group. However it is possible to show that *vg* is linked to other genes, which together constitute another linkage group. In *Drosophila*, very extensive recombinational analysis involving a very large number of genes has been carried out, and it has been shown that all genes are members of one of only four linkage groups. The linkage groups themselves are inherited independently of one another, so that genes in two different linkage groups show no linkage. The linkage groups are not all equal in size, that is they do not contain the same number of genes, nor do the sums of the individual recombination fractions of the member genes add up to give the same 'lengths'. In *Drosophila*, one of the linkage groups is, on both these criteria, much smaller than the other three.

Morgan also discovered another phenomenon related to linkage. He showed that certain characters in *Drosophila* were not inherited independently of sex. This type of inheritance is called **sex linkage**. An example of sex linkage is the inheritance of the white-eye mutation. If a female fly, from a stock which breeds true for the normal red eye colour, is crossed to a male from a true-breeding white-eyed stock, all the F_1 progeny are red eyed, and the F_2 progeny, obtained by crossing the F_1 progeny among themselves, show a 3 : 1 segregation for normal to white eyes. It can be concluded therefore that this difference in eye colour is the result of mutation in a single gene, and that the red-eyed allele is dominant to the white-eyed allele. It is when the F_2 progeny are classified for sex, that an anomaly emerges. It is found that the frequency of the two eye colour types is not the same in the two sexes. All the F_2 females are red eyed and of the males, half are red eyed and half white eyed. If the **reciprocal cross**, i.e. a female from the true-breeding white-eyed stock is crossed to a wild-type male, different results are obtained. In the F_1, all the females are red eyed, but all the males are white eyed. These results are summarised in figure 14.

The first cross established that the red-eyed (wild-type) allele was dominant to the white-eyed allele, and yet the second cross generated white-eyed males in the F_1. These males cannot therefore have inherited a red-eyed allele from their fathers. Conversely it may be concluded that since all the white-eyed flies among the F_1 progeny are male, not receiving the red-eyed allele from its father causes a zygote to develop into a male. The conclusion is therefore that males appear to have only one gene for eye colour, and that this is the cause

of their maleness. This is actually a little presumptuous. It could be
that the eye colour gene, and the gene causing maleness are alternatives. If this is so, the two crosses can be represented using gene
symbols as in figure 15, where + represents the wild-type allele, w the
white-eye colour allele, and M, the maleness factor. In *Drosophila*, it

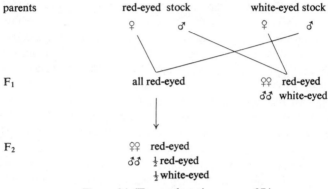

Figure 14. (For explanation, see p. 27.)

is found that one whole linkage group shows sex linkage, and so it
appears that the maleness factor, if it exists, would have to be the
equivalent of a whole linkage group. Sex linkage has been detected
in many organisms including man, where red–green colour blindness
and haemophilia are two well known characters showing sex-linked
inheritance.

Not all examples of sex linkage exactly parallel the *Drosophila*
situation. In the chicken, it appears that it is the female that has one
copy only of some genes, and that there might be a factor for femaleness. As a result the whole pattern of sex linkage is reversed. An
example of this is given in figure 16, where the inheritance of striped
or bar plumage (+) and non-striped or non-bar (*b*) plumage is shown.

We are now ready to equate what has been learnt about the nature
and behaviour of the genetic material in eukaryotic organisms from
breeding experiments, with the physical structure of the cell. Before
doing this however, the main conclusions which have been drawn from
breeding experiments will be listed. These are:

(1) Each cell contains either one or two copies of each gene. In
all organisms the gametes contain one copy, and the zygote two; but
in some organisms the adult cells contain two (e.g. flowering plants,

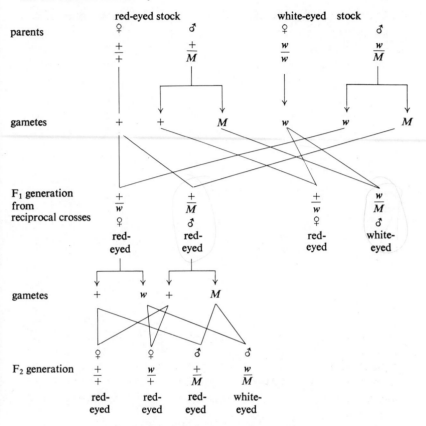

Figure 15. (For explanation, see opposite.)

and vertebrates) whilst in other organisms they contain one (e.g. fungi).

(2) There must be two methods of cell division, one involving the production of daughter cells containing the same number of gene copies (i.e. diploid from diploid, or haploid from haploid) and the other which results in the daughter cell having half the number of gene copies (i.e. reduction division, haploid from diploid).

(3) Throughout the life cycle the individual copies of the same gene behave in a particulate manner, not blending with one another.

(4) Although different genes may be inherited independently of one another, some combinations show linkage.

(5) The linkage values indicate that genes are associated together

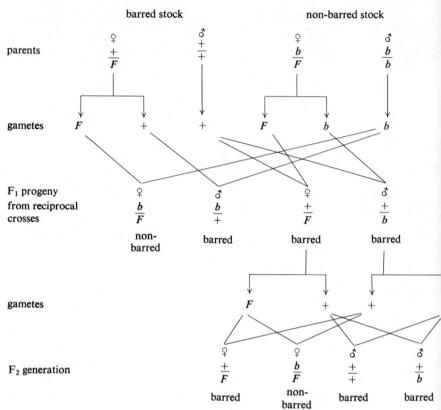

Figure 16. (For explanation, see p. 28.)

in discrete linkage groups, which are inherited independently of one another.

(6) The linkage values between genes in the same linkage group are consistent with their arrangement in a linear order.

(7) For any particular organism, there is a characteristic number of linkage groups, which may be of different sizes.

(8) In some organisms, the genes of one of the linkage groups show sex-linked inheritance. Depending on the breeding pattern it can be deduced which sex contains only one copy of the genes in the sex-linked linkage group.

All this can be deduced without looking at an individual cell. Observation of the cell immediately suggests that the chromosomes within the nucleus are the obvious candidates for the hereditary information. Circumstantial evidence favours the nucleus among cell organelles as the site of the hereditary information. Nearly all cells have a nucleus, and those that do not are specialised cells incapable of further division. Furthermore the gametes which fuse to form a zygote are often markedly different in size. But we have already deduced that they each contribute an equivalent amount of hereditary information to the zygote. If the two types of gametes are examined, it is found that the nuclei which they contain are more or less similar in size. However it is the observation of chromosome behaviour which confirms the hereditary role of the nucleus. This behaviour conforms exactly to the patterns that we have predicted as a result of breeding experiments. Although the chromosomes cannot be distinguished by cytological procedures between cell divisions, shortly before cell division, probably due to their shortening and thickening, they become visible under the light microscope when appropriate techniques are used. Once visible they can be seen to be rod-like structures of varying lengths. In a particular organism, cells which have been deduced to be diploid contain twice as many chromosomes as those which have been deduced to be haploid. Furthermore in diploid cells where chromosomes are morphologically distinguishable, they can be seen to occur in pairs of each type and for a given species it is found that the number of chromosomes in a haploid cell corresponds to the number of linkage groups established by breeding experiments. Finally in those species where sex-linked inheritance has been detected, it can often be seen that whereas in diploid cells of one sex one type of chromosome is present twice in the normal way, in the other sex this chromosome, called the X **chromosome**, is represented only once: it may have a morphologically distinct partner, which is then called the Y **chromosome**, or it may have no partner. This sex, because it produces gametes of two types, those with and those without an X chromosome, is said to be the **heterogametic sex**. The heterogametic sex is always the sex which breeding experiments indicate carries only one copy of the sex-linked linkage group, and so it appears that the Y chromosome must correspond to the proposed factor for maleness or femaleness.

In every respect so far, therefore, the chromosomes are candidates for the physical structures corresponding to linkage groups. However genes in a linkage group do not show complete linkage, but recombination occurs. Before seeing whether there is any aspect of chromosome behaviour which can be equated to the recombination process it is necessary to examine chromosome behaviour at cell division in detail.

The important thing to emerge is that the chromosomes show two quite distinct alternative types of behaviour. By far the most common type of cell division, called **mitosis**, involves the longitudinal division of each chromosome, and the distribution of one of the products of this division to each of the two daughter cells. By this means a diploid cell gives rise to diploid cells, or a haploid cell to haploid cells, and again the chromosomes are behaving as though they were the hereditary information. The second, rarer type of cell division, called **meiosis**, is more precisely two sequential divisions whereby a diploid cell gives rise to four daughter haploid cells, each of which contains only one of each of the chromosome pairs present in the diploid cells.

The process of mitosis is illustrated diagrammatically in figure 17 and in the following description the numbers in brackets refer to the stages illustrated in that figure. It should be remembered that the process is a continuous one, and that the diagrams are equivalent to individual stills chosen from a film. Between cell divisions, no chromosomes are visible in the cell nucleus, which has a more or less uniform granular appearance. (1) The onset of mitosis is marked by the chromosomes becoming visible as thin threads which continue to shorten and thicken (2) and (3). As this process proceeds each chromosome can be seen to be a longitudinally double structure, and to have a special region, often visible as a constriction, called a **centromere**. It can also be seen that where morphological differences occur between chromosomes there are two of each type. When the contraction process has finished, the chromosomes move to the centre of the cell, distributing themselves in the equatorial plane (4). At about this time, or sometimes a little earlier, the nuclear membrane breaks down. When the chromosomes have moved to the equator, it can be seen that the centromeres in particular are aligned in the same plane. Meanwhile the **spindle**, a structure radiating from the two poles of the cell, has been forming. This consists of fibres, and fibres from each pole become associated with the centromere of each chromosome. Then possibly as a result of a contraction of these fibres, or perhaps by some ratchet-like movement along them, the chromosomes are pulled in half along their longitudinal axis, one of each chromosome half (or **chromatid**) travelling to each pole (5). Next, new nuclear membranes are formed around the two groups of daughter chromo-

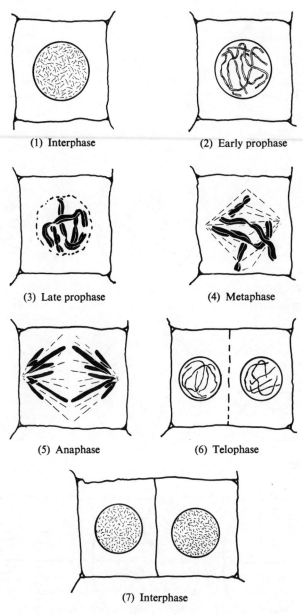

(1) Interphase

(2) Early prophase

(3) Late prophase

(4) Metaphase

(5) Anaphase

(6) Telophase

(7) Interphase

Figure 17. Schematic representation of the main stages of mitosis.

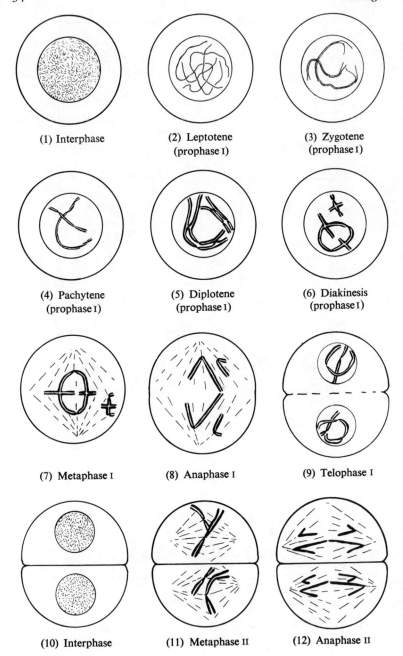

(1) Interphase

(2) Leptotene (prophase I)

(3) Zygotene (prophase I)

(4) Pachytene (prophase I)

(5) Diplotene (prophase I)

(6) Diakinesis (prophase I)

(7) Metaphase I

(8) Anaphase I

(9) Telophase I

(10) Interphase

(11) Metaphase II

(12) Anaphase II

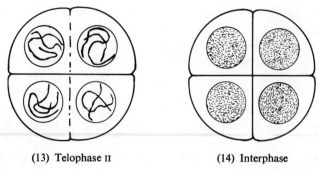

(13) Telophase II   (14) Interphase

Figure 18 (opposite and above). Schematic representation of the main states of meiosis.

somes, and so two nuclei are produced, each of which contains the same number of chromosomes as were in the original cell (6). Finally the cell itself divides either by the growth of a dividing wall, or by the nipping in two of the cytoplasm. Meanwhile the chromosomes become progressively less distinct in the two nuclei, which take on the typical between-divisions (**interphase**) appearance (7). It is during interphase that the chromosomal material is replicated, so that the single chromatid chromosomes which enter interphase consist of two chromatids by the time the cell enters the next mitosis.

Meiosis, which is illustrated diagrammatically in figure 18, is a more complex process, which is nevertheless clearly related to mitosis. The numbers in brackets refer to the stages illustrated in figure 18. The onset of meiosis is also heralded by the chromosomes in the diploid parent cell becoming distinct as a result of their shortening and thickening (2). They do not at first appear to consist of two chromatids, but it has nevertheless been shown by chemical means that chromosome replication has been substantially concluded by this time. Next the corresponding chromosome pairs come together and pair closely along their length (3). Sometimes pairing is so complete that the only obvious indication that it has occurred is that there are only half the number of these structures visible as there are chromosomes in the diploid cell (4). Chromosomes which pair at meiosis are said to be **homologous**, and the paired structure once formed is called a **bivalent**. As the process of shortening and thickening continues the homologous chromosomes in the bivalent lose their attraction for one another, and come apart. They are however prevented from coming apart completely, as they are held together at one or several points along their length (5). By this time the chromosomes can be seen to consist of two chromatids, and in especially favourable material, these points can be seen to be due to a reversal of the pairing relationships of the four chromatids involved in the bivalent (see Plate 1, page 37). Because the bivalent takes up a cross-like configuration around these points (6), they are called **chiasma** (plural **chiasmata**). Experiments can be performed which tell us more about the origin of chiasmata. Some of these will be described below. As in mitosis, the chromosomes then migrate to the cell's equatorial plane (7). The nuclear membrane meanwhile disappears, the spindle is formed, and the spindle fibres attach to the centromeres. What follows constitutes the crucial difference between meiosis and mitosis, for instead of half chromosomes (chromatids) migrating to each pole, so maintaining the chromosome number, one whole chromosome of each homologous pair migrates to each pole (8), and so the chromosome number of the two daughter cells is halved (9). Very soon after this first division of meiosis is concluded, another division occurs which is essentially similar to a mitosis (10–14). As a result four haploid cells, usually called a **tetrad,** are formed from the original diploid cell.

Returning to chiasmata, clearly if chromosomes are to be equated to linkage groups, some process is required which will allow genes on the same chromosome to recombine, otherwise linkage would be total. It is obviously attractive to suggest that chiasmata are the visible indications that this recombination process occurs. However several alternative theories to account for the origin of chiasmata have been put forward. The three principal hypotheses are:

(1) *Chromatid dissociation and reassociation.* On this hypothesis, it is envisaged that the two chromatids of a chromosome (= **sister chromatids**) which is involved in a bivalent, dissociate from one another. At a later stage reassociation occurs, but a chromatid can reassociate not only with its sister, but instead with one of the chromatids of the homologous chromosome so producing a chiasma (see figure 19).

(2) *Breakage and reunion.* This hypothesis proposes that chiasmata are produced as a result of the breakage at corresponding points of one chromatid of each of the homologous chromosomes involved in a bivalent, and their reunion one with the other (see figure 20).

(3) *Copy-choice.* The third hypothesis proposes that the second chromatid of each chromosome is replicated during the first meiotic

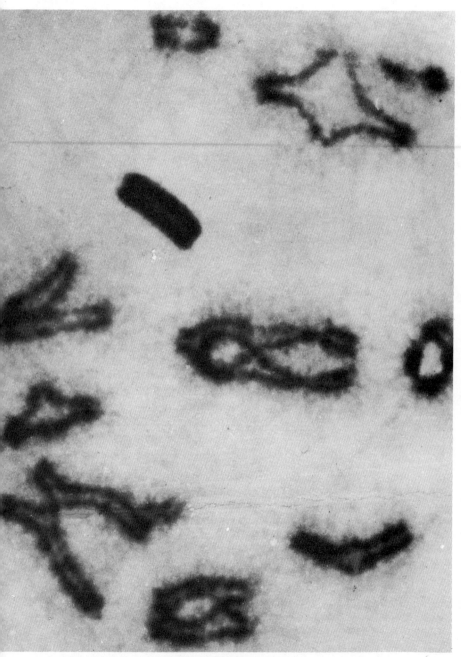

Plate 1. Diplotene stage of meiosis during sperm production in the locust,

prophase, after homologous chromosomes have paired. It further proposes that the two chromosomes provide alternative templates for the synthesis of the new chromatids, and that a chiasma is formed as a result of the new chromatids having switched their templates at that point (see figure 21).

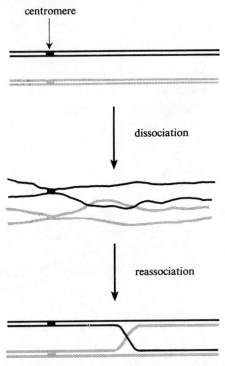

Figure 19. The chromatid dissociation and reassociation hypothesis for the origin of chiasmata.

If chiasmata are formed by chromatid dissociation and reassociation then they will not generate daughter chromosomes derived partly from each of the two homologous chromosomes. This is however a requirement if linkage groups are to correspond to chromosomes and genetic recombination is to be accounted for, which furthermore the two other hypotheses fulfil. It is possible to distinguish between the dissociation–reassociation hypothesis on the one hand, and the breakage–reunion and copy-choice hypotheses on the other. Rarely it is possible to obtain homologous chromosomes which can be distinguished one from the other. In a few cases the terminal section of one of the homologous chromosomes is missing. The predictions of what occurs if a chiasma is formed which involves this chromosome are shown diagrammatically in figure 22.

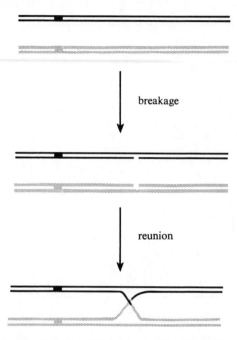

breakage

reunion

Figure 20. The breakage and reunion hypothesis for the origin of chiasmata.

The dissociation–reassociation hypothesis predicts that the homologous chromosomes would appear the same length if a chiasma occurs, whereas the other two hypotheses predict that they would remain morphologically distinct. All experimental observations are consistent with this latter prediction, and so the dissociation–re-association hypothesis can be rejected.

Proof that genetically observed recombinational events did correspond to chiasmata was provided by an elegant experiment by C. Stern, the results of which were published in 1931. Stern used *Drosophila melanogaster* for this experiment. In *Drosophila*, the male is the heterogametic sex, and the *Y* chromosome is readily distinguishable from the *X*. It is possible to obtain stocks which have

morphologically distinguishable *X* chromosomes. Stern used two
such stocks which will be referred to as *J* and *K*. Stock *J* had a broken
X chromosome. It was also true breeding for two mutations which
showed sex-linked inheritance, carnation, *car*, an allele affecting eye
colour which is recessive to the wild type allele, and bar, *B*, an allele

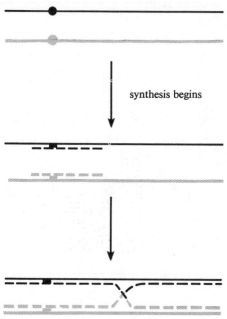

Figure 21. The copy-choice hypothesis for the origin of chiasmata.

affecting eye shape which is dominant to the wild type. Stock *K* was
a true-breeding wild-type stock, but had an *X* chromosome which had
an extra piece from one of the other chromosomes joined to it. Stern
crossed these two strains together and obtained an F_1 generation.
He took the females from this F_1 and back-crossed them to males from
a stock (stock *L*) which was true breeding for carnation eyes, and had a
normal *X* chromosome. The genetic and cytological constitution of
these stocks are represented diagrammatically in figure 23.

Stern concentrated on the females from the $F_1 \times L$ back cross.
Four classes of progeny were obtained in the usual way. These all
must have inherited an *X* chromosome of normal appearance, and a
carnation allele from their father. What they inherited from their

mother depended on whether recombination had occurred. By studying the chromosome complement of the four progeny classes cytologically, Stern was able to observe that those classes which were genetically parental for the characters involved had one X chromosome resembling either stock J or stock K. Thus the carnation, bar

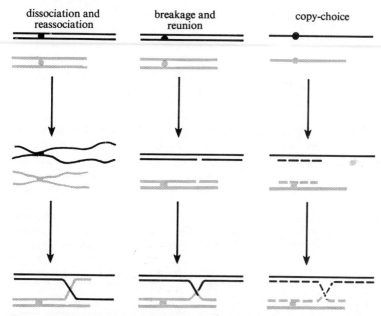

Figure 22. The outcome of chiasma formation between morphologically distinguishable homologous chromosomes predicted by the three hypotheses for the origin of chiasmata.

class of progeny, whose inheritance of these alleles can be traced back to stock J, had one broken X chromosome, and the wild-type class, one of whose X chromosomes had been derived from stock K, had one X chromosome with an extra piece joined to it. The crucial finding was that the genetically recombinant classes (carnation-coloured, normal-shaped eye; and normal-coloured, bar-shaped eye) also showed a recombinant configuration for one X chromosome. These findings are summarised in figure 24.

These results are therefore entirely consistent with genetically recognisable recombination corresponding to the cytologically observable event of chromosome exchange. They also indicate that the carnation and bar genes must be on the X chromosome between the break present in stock J, and the extra chromosome piece present in stock K. The structure of the chiasma necessary to produce the recombinant gametes in the F_1 must be as follows:

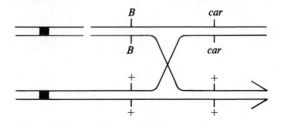

Stern's experiment, while showing the correspondence between chiasma and recombination, does not distinguish between the breakage–reunion and the copy-choice hypotheses of chiasma formation. Although the copy-choice hypothesis has many attractions,

	Sex	Eye shape	Eye colour	Genetical constitution	Cytological appearance of X chromosomes
Stock J	♀	bar	carnation	$\dfrac{B\ car}{B\ car}$	
	♂	bar	carnation	$\dfrac{B\ car}{M}$	
Stock K	♀	normal	normal	$\dfrac{+\ \ +}{+\ \ +}$	
	♂	normal	normal	$\dfrac{+\ \ +}{M}$	
F_1 (from $J\,K$)	♀	bar	normal	$\dfrac{B\ car}{+\ \ +}$	
Stock L	♂	normal	carnation	$\dfrac{+\ car}{M}$	

Figure 23. Details of stocks of *Drosophila melanogaster* used by C. Stern in his experiment to investigate the connection between chiasmata and genetic recombination.

and at one time had many followers, it now seems likely that it is not correct. Two only of the pieces of evidence which go against it and favour the breakage–reunion theory will be mentioned here, and these will be dealt with briefly. One has already been alluded to. It will be recalled that the chromosomes when they first became distinct at the

Eye shape	Eye colour	Genetical constitution	Recombination required?	Cytological appearance of X chromosomes
bar	carnation	$\dfrac{B\ car}{+\ car}$	No	
bar	normal	$\dfrac{+\ +}{+\ car}$	No	
normal	carnation	$\dfrac{+\ car}{+\ car}$	Yes	
bar	normal	$\dfrac{B\ +}{+\ car}$	Yes	

Figure 24. Details of the types of female progeny obtained from the cross between males from stock L and females from the F_1 between stocks J and K.

onset of meiosis, appear to be single structures, the individual chromatids only becoming apparent after chromosome pairing. This observation apparently favours the copy-choice hypothesis, which demands that chromosome replication occurs after chromosome pairing. However there is now chemical evidence that chromosome replication is largely completed before chromosome pairing occurs, a fact which makes the copy-choice hypothesis untenable. The other evidence against the copy-choice hypothesis is genetical. The copy-choice hypothesis demands that the replication of chromatids must be **semi-conservative**, that is to say the chromatids inherited by two of the daughter cells in the tetrad formed as a result of meiosis, must be entirely parental. Since recombination can only occur as a result of synthesis all recombinational events must be confined to the other two daughter cells. In some organisms, particularly fungi, it is possible to recover the four cells resulting from a single meiosis, and to analyse them genetically. This process, called **tetrad analysis**, has played a very important role in our understanding of the process of recombination. Suppose a cross was carried out between two fungal strains, differing with respect to three genes. These strains will be called $+++$ and $a\,b\,c$. Should a recombinational event occur in the same meiosis between both the a and b genes, and the b and c genes, on the copy-choice hypothesis, the four ascospores of the tetrad produced can only have genetic constitutions as follows: $+++$, $a\,b\,c$, $+\,b\,+$, $a\,+\,c$. However tetrads are found which are made up of spores of the following types: $+\,+\,c$, $a\,b\,+$, $+\,b\,c$, $a\,+\,+$, i.e. where recombination must have involved all four chromatids. This then is another observation inconsistent with the copy-choice hypothesis, and so it is probable that the origin of chiasmata is best described by the breakage–reunion hypothesis.

4. Gene interaction

So far in this book, genes have been dealt with as though they brought about their effects for the most part, in isolation from, and without being affected by, the environment or the other genes which an organism possesses. The phenomenon of dominance provides the one example which has been described where this is clearly not true. Here a strain having one dominant and one recessive allele is indistinguishable from a strain homozygous for the dominant allele, and so strains which have different genetic constitutions can appear similar. It is usual to refer to the genetic constitution of a strain as its **genotype**, and to its appearance as its **phenotype**. This distinction between genotype and phenotype which is largely the subject of this chapter, is of considerable importance.

The first example of how a strain of a given genotype can show different phenotypes in different environments, will be a simple one. It is possible to obtain mutant strains of *Aspergillus* which are unable to produce conidiospores at the normal culture temperature of 37 °C. If a number of independently-isolated sporeless strains are cultured at 25 °C, it is found that a small proportion are able to produce conidiospores at that temperature. Such strains which show this effect reproducibly, are said to carry a **temperature-sensitive** mutation. Temperature-sensitive mutations are perhaps the most straightforward example of how the environment can affect the phenotype. It is probable that the mutation has led, not to the absence of an enzyme, but to the production in the mutant strain of an enzyme which is more heat-labile than the wild-type product. At 25 °C this enzyme is active, but at 37 °C it is denatured and the strain therefore has a mutant phenotype. In *Aspergillus*, temperature-sensitive mutations are fairly common, and it is possible to obtain strains with, for example, temperature-sensitive biochemical mutations. Such strains require a biochemical to be added to the medium in order to grow at a high temperature, but at a low temperature they have no such requirement. There are also plenty of examples of differential environmental effects on strains of higher plants and animals. The biochemical basis of these is usually not so well understood, but an example is the size difference between Runner beans and dwarf French beans. In the garden in England, Runner beans characteristically grow to a height of about ten feet, dwarf French beans only grow to about one and a half feet. In a greenhouse however where temperatures are higher, the two varieties are indistinguishable, each growing to ten feet. Nor are environmental effects confined to temperature, it is for example possible to get strains of certain species of flowering plant which show marked differences in response to such environmental features as day length or soil type.

Where complex characters are involved, especially in organisms such as man where genetic experiments are impossible, it will be seen how difficult it will be to decide to what extent the phenotype of an individual depends on his genotype, or on the environment in which he developed. In fact it is meaningless to ask such a question, as no genotype can generate a phenotype without an environment, and *vice versa*. It is really how genotype and environment interact to produce the phenotype which is important, but this is a problem no less difficult to investigate.

The effects of mutations in other genes on the character determined by a particular gene can be thought of as extension of the effects of the environment. In some relatively simple cases, the bases of such interactions between genes are well understood. In fact the first example given in this chapter, that involving sporeless mutations in *Aspergillus* can be used to illustrate one type of gene interaction. If a sporeless strain is obtained from the wild type by mutation, and is crossed to a mutant strain producing yellow conidiospores, three classes of progeny only are obtained. Half the progeny are sporeless, and of the remainder, half have wild type green spores, and half yellow spores. This finding is perhaps not too surprising. If mutation in two different unlinked genes is responsible for the absence of spores and for the change of spore colour, then it is to be expected that four genetically distinct classes of progeny will be obtained in the normal manner. However it will be impossible to score conidiospore colour in a strain which does not produce conidiospores, and so two of those classes, the ones carrying the sporeless mutation will be indistinguishable, except by further breeding experiments. This experiment is represented diagrammatically in figure 25.

This is an example of the phenomenon of **epistasy**. The sporeless mutation is said to be **epistatic** to the spore colour mutation (or conversely the spore colour mutation is said to be **hypostatic** to the

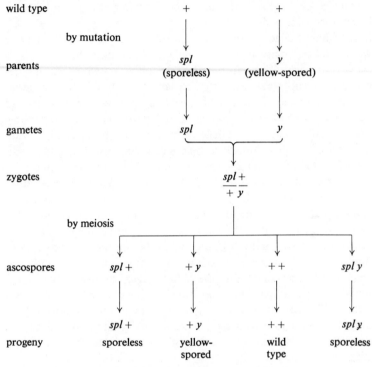

Figure 25. (For explanation, see opposite.)

sporeless mutation). Examples of epistatic interactions between genes are quite common. For example it is possible to get a recessive mutation in *Drosophila*, which, in strains carrying it homozygously leads to the absence of eyes. Such a mutation will obviously be epistatic to a mutation affecting eye shape, such as bar, or a mutation affecting eye colour.

Epistasy is also encountered when the inheritance of some mutations affecting more directly the same character is considered. The results of breeding experiments of this type can tell us a great deal about the basic effects of a gene mutation. As an example the inheritance of conidiospore colour in *Aspergillus* will once again be used.

It has been mentioned previously that as well as yellow-conidiospored mutants, it was possible to obtain strains of *Aspergillus* having chartreuse, pale green or white conidiospores. Crosses between such strains and a wild type indicate that each strain carries a single mutant gene. Since all affect the same character, it is possible that it is the same gene which is mutated in each strain to give alleles which have different effects. This can be tested by crossing the strains one with another, for example, if a pale green-conidiospored strain is crossed to a white-conidiospored strain, only the two parental types progeny are obtained, even though very large numbers of progeny may be examined. Such a result could be obtained if mutations had occurred in two extremely closely linked genes, but it is also possible that mutations in the same gene are responsible. In a haploid organism it is often very difficult to distinguish between these two possibilities, but ways of doing so will be discussed later in this chapter.

When a yellow-conidiospored strain is crossed to a chartreuse-conidiospored strain, the results obtained are more straightforward to interpret. Four classes of progeny are obtained in equal numbers, two of the classes have the parental spore colour, one has the wild-type green spore colour, and the fourth has spores of a new colour, light brown. It appears therefore that mutation in two unlinked genes must be responsible for the yellow and chartreuse spore colours, and that the new light-brown spored class must be the doubly mutant strains. This can be confirmed by taking one of this class of progeny and crossing it to a wild type, whereupon the same four classes of progeny are obtained in equal numbers. These crosses are shown diagrammatically in figure 26.

Exactly the same situation can occur in higher organisms. If a true-breeding red-flowered strain of *Streptocarpus*, the Cape primrose, is crossed to a true-breeding mauve-flowered strain, all the F_1 hybrids are blue. If these are crossed among themselves, it is found that $\frac{9}{16}$ of the resulting progeny are blue flowered, $\frac{3}{16}$ are red flowered, $\frac{3}{16}$ mauve flowered, and the remaining $\frac{1}{16}$ have salmon-coloured flowers. It can therefore be deduced that the mutations leading to the production of red flowers and of mauve flowers must be in two different unlinked genes, and each mutant allele must be recessive to the wild type which has blue flowers. Strains homozygous for both mutations, which will account for $\frac{1}{16}$ of the F_2, have salmon flowers. This can be confirmed by taking such strains and crossing them to a blue-flowered, wild-type strain, when a similar pattern of inheritance

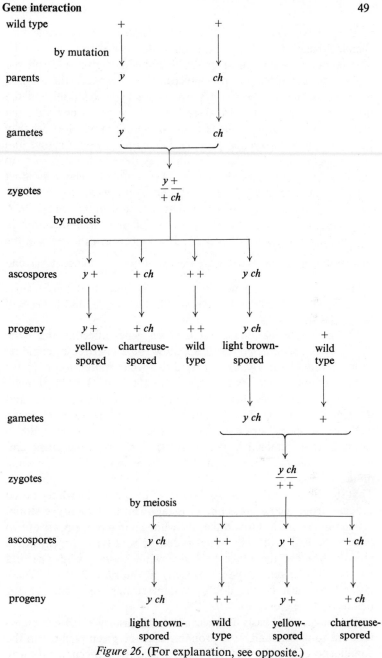

Figure 26. (For explanation, see opposite.)

to the red with mauve cross will be obtained. The probable bio-chemical basis of this situation is understood; it is thought that the blue colour is due to the pigment delphinidin diglycoside and the salmon to pelargonidin monoglycoside. To convert the salmon pigment to blue requires two steps, one a hydroxylation, and the other the addition of a sugar residue. These two steps are each catalysed by an enzyme, and it does not matter in which order they proceed. The hydroxylated product, having no sugar, is mauve and so it appears that the mauve strain lacks the enzyme necessary to catalyse the addition of the sugar. The non-hydroxylated product with the sugar residue added is red, and so it appears to be the en-zyme catalysing the hydroxylation that is missing in the red strain. A very simplified scheme for the pathway of pigment production in *Streptocarpus* is given in figure 27. The F_1 between the red and the mauve-flowered strains will have the genotype $\dfrac{r\ +}{+\ m}$. Since it has one wild-type copy of each gene, it will be able to make both the enzymes involved in the synthesis of delphinidin diglycoside, and hence will have blue flowers.

The next type of interaction between genes affecting the same character is encountered when a yellow (or a chartreuse) conidio-spored strain of *Aspergillus* is crossed to a white-spored strain. If this is done, only three classes of progeny are obtained. Half the progeny are white-spored, one quarter green-spored like the wild type, and the remaining quarter have yellow spores (or chartreuse spores when a chartreuse-spored strain was used as a parent). This result is remin-iscent of that obtained in the sporeless × yellow-spored cross, and would be explained if the mutated genes were unlinked, and the white mutation was epistatic to the yellow (and chartreuse) mutations. That this is indeed the case can be shown by taking the white-spored progeny from such a cross and crossing them to a wild-type strain. When this is done it is found that the white-spored progeny are of two types: one half give only white-spored or wild-type progeny when crossed whereas the other half give white-spored, wild-type and yellow (or chartreuse) spored progeny in the ratio 2 : 1 : 1. These latter type must be the doubly-mutant strains (see figure 28 for details of these crosses).

The biochemical basis of this example of epistasy is again straight-forward to understand. It is probable that the green pigment in the conidiospores is manufactured from a colourless precursor by way of a yellow intermediate as follows:

$$\text{colourless precursor} \xrightarrow{1} \text{yellow intermediate} \xrightarrow{2} \text{green pigment.}$$

These conversions are enzyme catalysed. If an enzyme necessary to convert the yellow intermediate to the green pigment is missing the spores will be yellow, and if an enzyme necessary to convert the colourless precursor to the yellow intermediate is missing the spores will be colourless (white). The spores will also be white if both enzymes are missing. The pattern of inheritance in the cross between a

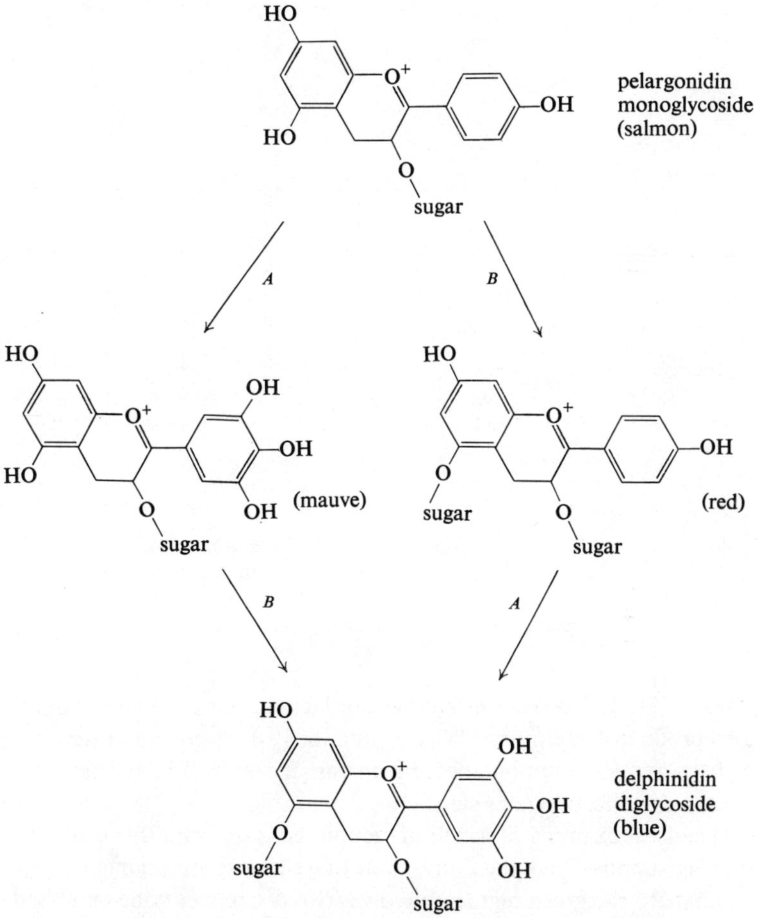

Figure 27. Simplified scheme of flower pigment synthesis in *Streptocarpus*. Enzyme *A* catalyses the hydroxylation step, and enzyme *B* the addition of a sugar residue.

yellow-conidiospored and a white-conidiospored strain is consistent with the yellow and white alleles leading to the absence of enzymes which act sequentially in the same pathway. The white strain would be unable to catalyse step 1, and the yellow strain, step 2 in the scheme

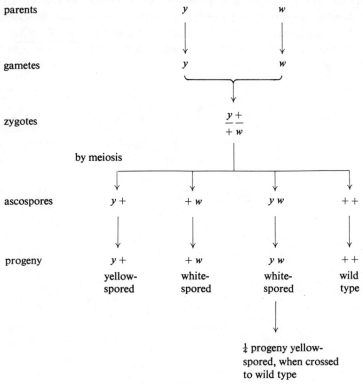

Figure 28. (For explanation, see p. 50.)

on page 51. The doubly mutant strain lacking both enzymes would also produce white spores. Where mutations affect sequential steps in a pathway, the mutation affecting an earlier step will be epistatic to mutations affecting later steps.

The final example of gene interaction leads on from this example. Let us suppose that the conversion of yellow spore pigment intermediate to the green pigment involves two discrete enzyme-catalysed steps and that the product of the first step is also yellow coloured, visually indistinguishable from the original yellow intermediate.

The scheme would now be:

$$\text{colourless precursor} \xrightarrow{\ 1\ } \underset{A}{\text{yellow intermediate}} \xrightarrow{\ 2\ } \underset{B}{\text{yellow intermediate}} \xrightarrow{\ 3\ } \text{green pigment.}$$

Strains carrying a mutation leading to an absence of the enzyme catalysing step 2, or step 3 will be indistinguishable both producing yellow-coloured conidiospores. Only upon crossing would it become evident that two genes were involved. It is customary in many organisms to designate the same symbols to mutations having the same effects, but to indicate that different genes are involved by adding a capital letter after the symbol, and thus in this example the two mutations will be called y-A and y-B. The predicted breeding scheme is shown in figure 29.

The doubly-mutant strain will also produce yellow conidiospores, and so if the two genes involved are unlinked, it is expected that the ratio of yellow-spored to green-spored progeny will be 3 : 1. This is the first time a 3 : 1 ratio has been encountered in a haploid organism. Situations of this sort are common in *Aspergillus*, and for example a histidine requirement can be caused by mutation in any of ten genes.

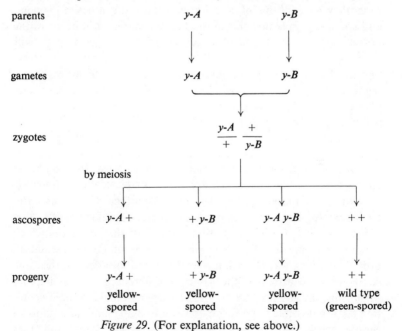

Figure 29. (For explanation, see above.)

The detection of wild-type recombinants as a test of whether two genes are involved, runs into difficulties if those two genes are linked. If the genes are closely linked, the recombinant wild-type class may be so reduced in frequency that they are difficult to detect. There are in fact other methods which can be used to detect if two genes are involved, which will be described below. Firstly however, the corresponding situation in a diploid organism will be examined.

Several different varieties of white-flowered sweet pea exist, and the result of crossing two of these together will be considered. When the two varieties 'Emily Henderson' and 'Blanche Burpee,' both of which are true breeding for white flowers, are crossed together, the resulting F_1, have purple flowers. These give rise to F_2 progeny which show segregation for purple or white flowers in the ratio of 9 : 7. This is consistent with two unlinked genes being involved, either one of which can undergo mutation leading to the absence of purple pigment synthesis, and hence to the production of white flowers. If the mutant alleles at the genes concerned are designated w-A and w-B, the breeding scheme shown in figure 30, is expected. Considering the F_1 generation first. Since these possess one wild-type allele of each gene, they have all the information necessary to synthesise pigment, and so have purple flowers. In the F_2, segregation will occur in the normal manner (cf. figure 10, page 18), and $\frac{9}{16}$ of the progeny will inherit at least one wild-type allele for each gene. These will, therefore, like the F_1, possess all the information necessary to make purple pigment. The remaining $\frac{7}{16}$ will be homozygous for the mutant alleles of one or both of the two genes, and so being unable to make pigment, will have white flowers.

When mutation in different genes can lead to the same effect, the genes are said to be **complementary**. It is usual to call the production of a wild-type product in, for example, the F_1 between two strains having mutations in complementary genes, **complementation,** and also to say that the mutations **complement** one another. This latter statement is to some extent an example of muddled thinking as it can be argued that it is really the wild-type alleles of the two genes which complement one another. The phenomenon of complementation provides a quick way of testing whether mutations affecting the same character have arisen in the same or different genes. In haploid organisms, the absence of a diploid phase would make complementation tests impossible to perform, however in many haploid organisms the life cycle is more complex than has so far been indicated, and other phases exist which enable complementation tests to be made. In *Aspergillus* for example, there are both **heterokaryon** and diploid phases. The cells of the mycelium of *Aspergillus* are multinucleate.

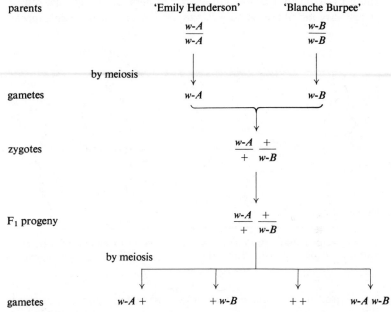

If the *w-A* and *w-B* genes are unlinked, the four types of gamete will be produced in equal numbers. These gametes will fuse randomly to generate an F_2 generation as follows:

$\frac{9}{16}$ having at least one wild-type allele of each gene

$\frac{3}{16}$ homozygous for the *w-A* gene

$\frac{3}{16}$ homozygous for the *w-B* gene

$\frac{1}{16}$ homozygous for both the *w-A* and the *w-B* gene.

Figure 30. (For explanation, see opposite.)

The mycelium of two strains of *Aspergillus* grown in close proximity will fuse to form a loose association in which the nuclei of the two strains share a common cytoplasm. This association, called a heterokaryon, provides a way for testing for complementation in *Aspergillus*. Suppose we obtain two biochemical mutants each of which requires, for example, tryptophan for growth. It is possible that these strains carry mutations affecting the same enzyme or two sequential enzymes in the biosynthesis of tryptophan. If the latter is true then a heterokaryon between them will possess wild-type alleles for both genes, and so will be able to make both enzymes, and synthesise tryptophan for itself. This situation is depicted diagrammatically in figure 31.

In *Aspergillus*, it is also possible to obtain a diploid phase, although this is by no means true of all fungi. *Neurospora* for example has only a heterokaryotic phase. *Aspergillus* mycelium produces conidiospores

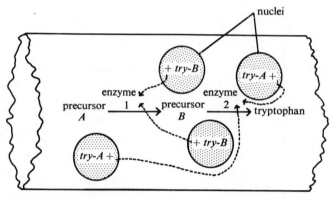

Figure 31. Schematic representation of a piece of mycelium of a hetero-karyon between two complementing tryptophan-requiring mutant strains (*try-A* + and + *try-B*).

which are for the most part haploid, and which upon germination will give rise to one or other of the two component haploid strains. However, a small proportion of the conidiospores produced by a heterokaryon are diploid, combining the genomes of the two original haploid strains. Such diploid conidiospores can be selected because of the phenomenon of complementation. Suppose the two component strains of a heterokaryon carried biochemical mutations leading to requirements for the vitamins biotin and riboflavin respectively. The heterokaryon would be able to grow without the addition of these vitamins to the medium. However the majority of the conidiospores it produced, being haploid, would require either biotin or riboflavin to germinate and grow. Only the diploid conidiospores combining the two genomes could grow on unsupplemented medium, so providing a selective technique for obtaining the diploid phase. The diploid phase of *Aspergillus* is not very stable and it breaks down, either spontaneously or by certain treatments, to give haploid strains. The mechanism of this process is not understood but it is probable that one of a pair of homologous chromosomes is lost first, producing a strain which is known as an **aneuploid**. Aneuploid strains are even more unstable, and so successive chromosome loss ensues until the haploid number of chromosomes is restored. This process provides an opportunity for reassortment of chromosomes, but no opportunity for recombination to occur. Because of this, genes in the same linkage

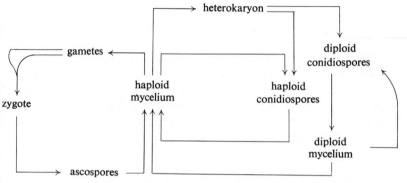

Figure 32. Life cycle of *Aspergillus*.

group will show complete linkage, a fact which is exploited for the genetic analysis of *Aspergillus*. The full life cycle of *Aspergillus* is given in figure 32.

Other groups of haploid organisms have stages akin to the heterokaryotic phase, which can be used for making complementation tests. Methods of performing complementation tests in certain bacteria and viruses will be described later in this book.

During the course of this book, experiments have been described which may be used to provide operational definitions of the gene. Thus a gene could be defined as the region of the hereditary material in which mutations occur which show no recombination with one another. Alternatively, a gene could be defined by the complementation test, on the basis that mutations in the same gene will not complement with one another. In fact these two definitions are often consistent with one another, however exceptions occur, and particularly the first definition could prove unreliable. The basis of these exceptions is now often understood but our understanding has usually depended upon genetic analysis of bacteria and viruses, two groups which have not so far been referred to. Before considering the genetics of these organisms however, consideration will be given to the chemical nature of the genetic material and the mutation process.

Problems

Problem solving is one of the most rigorous ways of ensuring that you understand the concepts of genetics which have been introduced in the first four chapters of this book. A few problems are given below which test your powers of logical deduction. The answers to the problems are given on page 188. There is no need to invoke mutation in any of your explanations.

1. In man:

(i) the ABO blood group is determined by a single gene with three alleles I^A, I^B and I^O. Individuals with blood group A have the genotype $\frac{I^A}{I^A}$ or $\frac{I^A}{I^O}$; with blood group B have the genotype $\frac{I^B}{I^B}$ or $\frac{I^B}{I^O}$; with blood group AB have the genotype $\frac{I^A}{I^B}$; and with blood group O have the genotype $\frac{I^O}{I^O}$.

(ii) the *MN* blood group is determined by a single gene with two alleles, *m* and *n*. Individuals with blood group *M* have the genotype $\frac{m}{m}$; with the blood group *N* have the genotype $\frac{n}{n}$; and with blood group *MN* have the genotype $\frac{m}{n}$.

(iii) the ability to taste phenyl thio-carbamide (PTC) depends upon the possession of a dominant allele *T*. Individuals homozygous for the recessive allele *t*, are unable to taste this chemical.

Solve the following problems.

(*a*) A woman whose blood group is O has four children. Their blood groups are A, A, B and O. The woman has only been married once, how many of the children must be illegitimate? ONE

(*b*) A woman whose blood group is B, and is unable to taste PTC, had three children. The first had blood group A, and could taste PTC, the second had blood group B and could taste PTC, and the

58

third had blood group O and was unable to taste PTC. What can you say about the genotype of the woman and her husband?

(*c*) A mother in a maternity ward claimed that the baby (*X*) allocated to her was not her own. There was only one other baby (*Y*) in the ward at the time. The mother has blood groups O, *MN* and cannot taste PTC. Baby *X* has blood groups A, *M* and can taste PTC, and baby (*Y*) has blood groups O, *MN* and cannot taste PTC. The mother's husband is dead, but she has three other children as follows:

    (i) blood groups A, *MN*,   able to taste PTC,   OO MN tt
    (ii) blood groups B, *N*,   able to taste PTC,   AB MN Tt
    (iii) blood groups B, *M*,   unable to taste PTC.

Which of babies *X* and *Y* is hers?

(*d*) A farmer had two sons. The first, born when he was young, grew into a handsome, healthy youth in whom he took great pride. The second, born much later, was always a sickly child and neighbours' gossip induced the farmer to bring his wife to court disputing its paternity. The grounds of the dispute were that the farmer, having produced so fit a first son, could not be the father of the weakling. The blood groups of the involved parties were as follows:

    farmer   O, *M*
    farmer's wife AB, *N*   the first son
    first son   A, *N*   was not his.
    second son   B, *MN*.

What advice would you give the court?

2. A flowering plant (i.e. a diploid organism) of unknown genotype was crossed to a plant from a strain which bred true for the characters colourless seed coat and round seed shape. The cross gave the following progeny:

    red, round seed   29   1 : 1
    colourless, square seed   31
    colourless, round seed   2438
    red, square seed   2502.

What was the genotype of the first plant, and what is the recombination frequency between the genes involved?

3. A female *Drosophila* with grey wings was crossed to a male with grey wings. The offspring were 114 grey-winged females, 53 grey-winged males and 58 yellow-winged males. What was the genetic constitutions of the original female?

4. Data from *Aspergillus* crosses involving the four linked genes *paba*, *pro*, *bi* and *y* were as follows:

Cross (*a*) *paba* × *y pro*

Progeny

paba y pro	2	*paba* + +	164
paba y +	27	+ + *pro*	29
paba + *pro*	16	+ *y* +	18
+ *y pro*	156	+ + +	4

Cross (*b*) *paba* × *bi y*

Progeny

paba bi y	55	*paba* + +	332
paba + *y*	3	+ *bi* +	4
paba bi +	19	+ + *y*	22
+ *bi y*	321	+ + +	61.

(*a*) In what order do the genes lie in the linkage group?
(*b*) Calculate the recombination fractions of the genes and draw a map of the linkage group.

5. Two true-breeding strains of maize, one of which was dwarf with crinkly leaves and brown stems, and the other of which was tall with normal leaves and green stems, were crossed. The F_1 plants were all tall with normal leaves and green stems. These F_1 plants were back-crossed to the dwarf, crinkly-leaved, brown-stemmed strain and gave offspring as follows:

tall, normal leaves, brown stems	201
dwarf, normal leaves, green stems	48
dwarf, crinkly leaves, green stems	196
dwarf, normal leaves, brown stems	43
tall, normal leaves, green stems	212
tall, crinkly leaves, brown stems	50
tall, crinkly leaves, green stems	46
dwarf, crinkly leaves, brown stems	206.

What can you say about the linkage of the genes concerned?

6. All the F_1 progeny from both reciprocal crosses between the two true-breeding strains of an insect, one with white eyes, and the other with normal green eyes, had green eyes. F_1 females were back-crossed to males from the white eyed strain and gave the following offspring:

green eyed 121, blue eyed 116, yellow eyed 127, white eyed 123.
How do you account for these results?

7. A cross was made between two true-breeding strains of sweet pea, one with red flowers and the other with white. The F_1 progeny all had purple flowers, and the F_2 progeny obtained from this F_1 were as follows:

   ·purple flowered 1192, red flowered 386, white flowered 496.

How do you account for this pattern of inheritance? Suggest a biochemical basis for these results.

8. Two true-breeding strains of plant both with purple flowers were crossed. The F_1 progeny all had purple flowers, but among 436 F_2 progeny obtained from them, there were 27 with white flowers, the remainder being purple flowered.

Suggest a possible explanation for these results.

9. A purple-flowered, hairy-leaved plant from a true-breeding stock was crossed to a white-flowered plant having no hairs on its leaves, also from a true-breeding stock. The F_1 progeny all had purple flowers and hairy leaves, and were back-crossed to the white-flowered, hairless strain. The resulting progeny were as in the table.

Flower colour:	Purple	Red	Blue	White	Total
Hairy leaves	362	204	196	369	1131
Hairless leaves	128	68	73	133	402
Total	490	272	269	502	1533

What can you say about the genetic differences between the two true-breeding stocks?

10. Two plants each from a different stock which bred true for white flowers, were crossed. Of the F_1 progeny approximately $\frac{3}{4}$ had white flowers and the rest had red flowers. The red-flowered F_1 plants were selfed, and of the F_2 progeny $\frac{9}{16}$ had red flowers and the rest had white flowers.

Account for these results, and predict the outcome of all the possible crosses between the F_1 plants.

11. Four females from a *Drosophila* stock, which was true-breeding for the character vestigial wings, were mated to males from a wild-type stock, each pair of flies being reared separately. The F_1 progeny

were examined, and in some cases were used to breed an F_2 generation. The results obtained were as follows:

First pair	35 wild-type females	219 wild-type females
	38 wild-type males	80 vestigial-winged females
		230 wild-type males
		73 vestigial-winged males
Second pair	38 wild-type females	Not grown
	34 miniature-winged males	
Third pair	54 wild-type females	Not grown
	28 wild-type males	
	23 miniature-winged males	
Fourth pair	43 wild-type females	243 wild-type females
	50 miniature-winged males	238 miniature-winged females
		152 vestigial-winged females
		230 wild-type males
		243 miniature-winged males
		170 vestigial-winged males

Explain these results.

12. True-breeding white-flowered strains, A, B, C, D and E, of a plant species were crossed in all possible ways. Each F_1 was uniform. Their flower colours are given below. What can you say of the genetic constitutions of the five strains?

		Strains used as ♀ parent				
		A	B	C	D	E
Strains used as ♂ parent	A	W	W	M	M	W
	B	W	W	M	R	W
	C	M	M	W	W	W
	D	M	R	W	W	W
	E	W	W	W	W	W

M = Mauve; R = Red; W = White

13. A green-conidiospored strain of *Aspergillus* was crossed to a yellow-conidiospored strain, and the following progeny were

obtained:

  56 green-spored, requiring adenine
  54 green-spored, with no requirement for adenine
 108 yellow-spored, with no requirement for adenine
  2 yellow-spored, requiring adenine.

Neither of the original strains required adenine. What might be the genetic explanation of these results? Suggest the possible biochemical basis of your explanation.

14. Females from a true-breeding white-eyed stock of *Drosophila* were mated to males from another true-breeding, white-eyed stock. All the F_1 progeny had wild-type dark-red eyes. The F_2 progeny obtained from them were as follows:

	Females	Males
Dark-red eyes	381	188
Bright-red eyes	129	62
Brown eyes	123	66
White eyes	40	358

How do you account for these findings? Would you expect the reciprocal F_1, and the F_2 obtained from it to show different results, and if so what would these be?

5. The chemical nature of the hereditary material and of mutation

The discovery which led to the series of experiments which were to provide the first direct evidence of the chemical nature of the hereditary information, was reported by F. Griffith in 1928. Griffith studied the bacterium *Diplococcus pneumoniae*, responsible for pneumonia in mice. Normal strains of *Diplococcus* owe their virulence to the possession of a polysaccharide coat, which renders them resistant to their host's defence mechanisms. These virulent strains, are also known as smooth, because they form smooth, shining colonies when grown on medium in a petri-dish. It is possible to obtain from the virulent strain, mutant non-virulent strains which have no polysaccharide coat. These strains can also be recognised because they form, on petri-dishes, colonies which have a rough, matt appearance. Griffith showed that no infection occurred in mice which had been injected with these non-virulent strains, nor were they infected if they were injected with virulent strains which had been killed by heat. However, if mice were injected with a mixture of a non-virulent strain and a heat-killed virulent strain, then a large proportion of the mice died, and moreover, it was possible to isolate from the dead mice, living virulent bacteria. Griffith concluded that the non-virulent bacteria had received something from the dead virulent bacteria which had allowed them once again to synthesise a polysaccharide coat, and he called this agent, a **transforming principle**. He speculated on the chemical nature of the transforming principle, suggesting that it might be the coat polysaccharide itself, which perhaps had an autocatalytic activity in its own synthesis. It was other workers who elucidated the true nature of the transforming principle, but this process took until 1944. Important steps were the discovery that the **transformation** of non-virulent strains to virulent could occur *in vitro*, that cell to cell contact was unnecessary, and that non-virulent strains could be transformed by the medium in which virulent strains had been growing. In 1944, O. T. Avery, C. M.

64

Macleod, and M. McCarty published the results of experiments which showed that transformation could be caused by highly purified **deoxyribose nucleic acid, DNA.** Furthermore the transforming principle was resistant to enzymes which hydrolysed protein or the other nucleic acid, **ribonucleic acid, RNA,** but was destroyed by DNAase, an enzyme which specifically attacks DNA. Even then the full significance of these findings were not appreciated, and it was suggested that DNA was acting as a selective mutagen, converting the non-virulent strain back to virulence, a step which can occur spontaneously but only with an extremely low frequency.

That DNA was itself the hereditary material was a fact which, despite the evidence of transformation, and of other experiments to be described, escaped most scientists until its structure was proposed by J. D. Watson and F. H. C. Crick in 1953. There had meanwhile accumulated a body of evidence, much admittedly circumstantial, that the hereditary material was DNA.

The evidence that, in eukaryotic organisms, the chromosomes contain the genetic information, has already been reviewed. It can be shown using stains which specifically combine with DNA, that the DNA in a cell is a major constituent of the chromosomes, and also confined to them. The other major constituent of chromosomes is protein, which is not of course confined in its distribution to the chromosomes; however, because many different types of protein molecule were known to exist, it seemed possible that some of these could have a hereditary role. In fact because the structure of proteins was known sometime before the structure of DNA, and because this structure, a linear polymer of various different amino acids which could be arranged in an enormous variety of different orders, seemed ideal for the storage of hereditary information in an encoded form, protein was the favoured candidate. DNA, whose chemical composition appeared too constant for it to function as an information store, was thought to have a structural function. It now seems possible that this is, at least in part, the role of the chromosomal proteins.

Some stains combine with DNA quantitatively and can be used as a direct assay of the DNA content of a nucleus. When this is measured it is found that the DNA behaves exactly as would be predicted were it the hereditary material. Thus the DNA content of a daughter cell is exactly half that of a cell about to undergo mitosis. The DNA content of pre-mitotic nuclei from a wide range of tissues is constant, except that haploid nuclei contain half as much DNA as diploid nuclei. Other circumstantial evidence, favouring DNA's genetic role, includes its relative stability. Compared with protein or carbohydrate, DNA turns over comparatively slowly in the cell. Finally there is mutagenic evidence. Perhaps the most direct is the effect of ultraviolet irradiation in causing mutation. Ultraviolet light is not equally mutagenic at all wavelengths, and the wavelength of 254 nanometres has the most effect. This wavelength corresponds to that which is maximally absorbed by nucleic acids. The absorption peak of proteins is however near to 280 nanometres, a wavelength which is a good deal less effective at causing mutation. Other mutagenic evidence is provided by treatment with chemicals that react specifically with DNA, and also by the feeding of nucleic-acid analogues–chemicals which differ only slightly from the bases normally found in DNA (see below, page 68). Both types of treatment are extremely effective at causing mutation, and thus provide further evidence of DNA's hereditary role. It has however been found that amino-acid analogues also cause mutation, but it is believed that this is a secondary effect. It seems probable that the amino-acid analogues are incorporated into enzyme proteins, which then no longer function as reliably as before. Thus the enzymes responsible for DNA synthesis are affected, and mistakes are made in DNA replication.

Other direct evidence that DNA, not protein, is the hereditary material, has been provided by experiments on viruses, performed during the 1950s. Two of the most important will be mentioned here. In 1952, A. D. Hershey and M. Chase, published the results of an experiment using a virus which parasitised a bacteria. Bacterial viruses, usually called **bacteriophages**, or 'phages for short, consist of protein and nucleic acid, in some types the nucleic acid is DNA, and in others RNA, but both types of nucleic acid are not found in the same virus. Hershey and Chase labelled viruses using either the radioactive isotope of sulphur, ^{35}S or of phosphorus ^{32}P. As DNA contains no sulphur, and protein no phosphorus the method provides a way of labelling these components specifically. They used the labelled virus particles to infect unlabelled bacteria, and then determined the distribution of radioactive isotopes. When infected with ^{32}P-labelled virus, they found that all the radioactivity entered the bacteria, but when ^{35}S-labelled virus was used, the radioactivity remained outside the bacteria, attached to the bacterial coat. They argued that it must be the DNA alone that entered the bacterium, and that this must contain the information on how to synthesise new virus particles, complete with their protein component.

The second important experiment using viruses, used a plant virus, the tobacco mosaic virus, usually abbreviated to T.M.V. This virus, in common with all plant viruses so far characterised, consists only of RNA and protein. The experiment, published by H. Fraenkel-Conrat and B. Singer in 1957, relied on the finding that it was possible to separate T.M.V. into its two components, protein and RNA, and to reaggregate them *in vitro* to give viable virus particles. It is also possible to obtain mutant strains of T.M.V., which can be recognised because they show a different pattern of infection in the host plant. Fraenkel-Conrat and Singer constructed *in vitro*, hybrids in which the protein was derived from one of these strains, and the nucleic acid from another. They found that these hybrids always showed the pattern of infection which was characteristic of the strain from which the nucleic acid of the hybrid had been derived.

Having reviewed the evidence in favour of DNA being the genetic material, it is now necessary to consider its chemical structure. If DNA is hydrolysed, it can be shown to consist of three quite different types of molecule. These are phosphate residues, the pentose sugar 2-deoxyribose, and nitrogen-containing bases. These bases are of four kinds, two belonging to the class of compounds called **purines**, and two to the class called **pyrimidines**. The two purines are **adenine** and **guanine**, and the two pyrimidines are **thymine** and **cytosine** (see figure 33 for details of the chemical structures). Milder hydrolysis of DNA yields **nucleotides**, which are compounds composed of a base, one deoxyribose residue, and a phosphate (the structure of the nucleotide which contains adenine, 3'-deoxyadenylic acid, is also given in figure 33). The nucleotides are linked together in DNA in chains by the phosphate residues which join the carbon atoms in the 3' position of one deoxyribose molecule to the 5' positions of the next deoxyribose molecule. Because the nucleotides are linked by the phosphate residues asymmetrically 3' to 5', a polynucleotide chain has polarity. Taking the simplest example, a dinucleotide between for example adenine and guanine, two possibilities exist: adenine–sugar–3'–phosphate–5'–sugar–guanine and guanine–sugar–3'–phosphate–5'–sugar–adenine. These are shown in figure 34. A polynucleotide chain therefore has polarity.

Several other discoveries about the chemical nature of DNA led to the final elucidation of its structure. It was found that the ratio of the bases obtained when DNA was hydrolysed was not haphazard. Firstly the total molar yield of purine obtained always equalled the molar yield of pyrimidine. The relationship was shown to be more

Purines

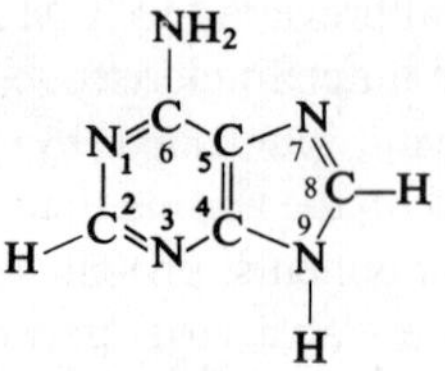

adenine

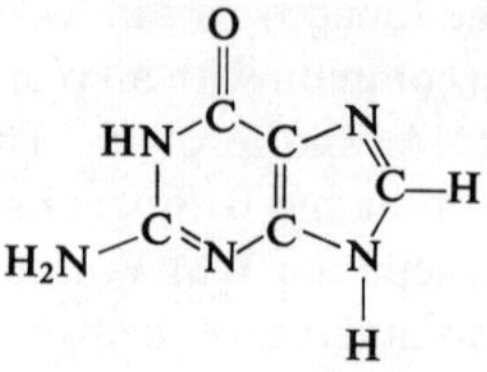

guanine

Pyrimidines

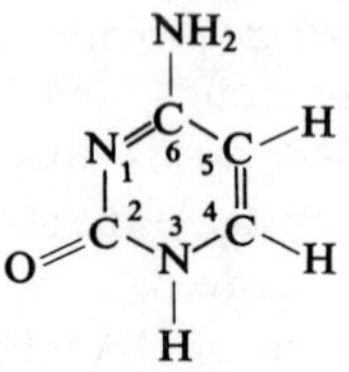

cytosine

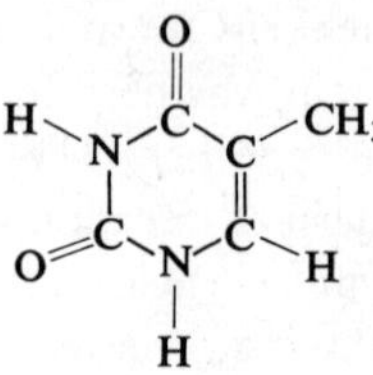

thymine

Sugar

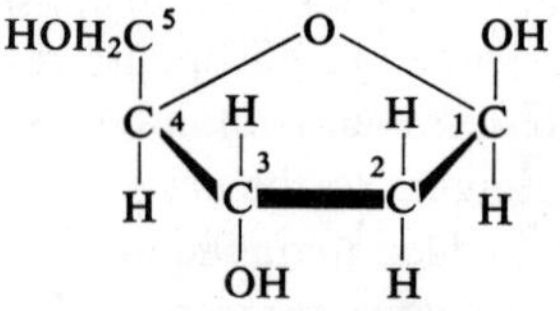

2-deoxyribose

Nucleotide

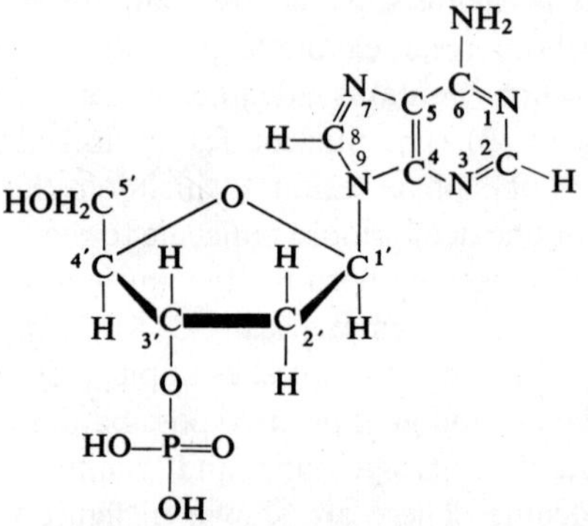

3′-deoxyadenylic acid
(the 3′ refers to the position of the phosphate residue)

Figure 33. Components of deoxyribosenucleic acid.

precise than this however. E. Chargaff showed that there were equal molar quantities of adenine and thymine contained in DNA, and also equal quantities of guanine and cytosine. Finally it was shown that the DNA extracted from a particular species showed a characteristic ratio of (adenine + thymine) to (guanine + cytosine), but that this ratio, usually called the **base ratio** could vary quite widely between species.

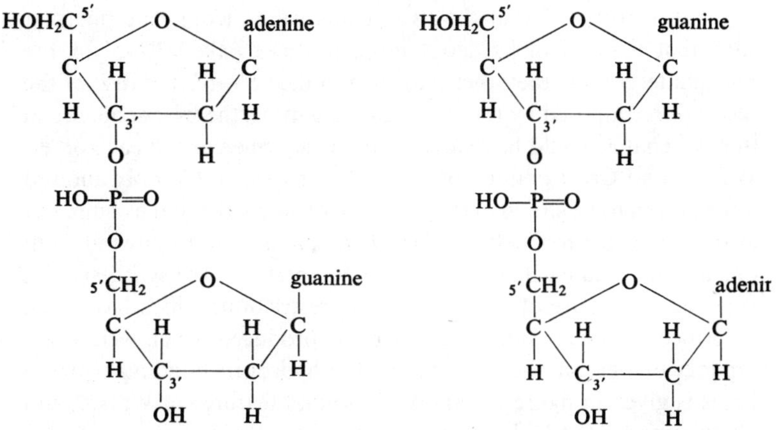

Figure 34. (For explanation, see p. 67.)

In addition to this chemical information about the structure of DNA, various physical characteristics were deduced by M. H. F. Wilkins and his co-workers. Using X-ray diffraction analysis, they discovered that DNAs from various sources all had basically the same structure, with repeating units at 340 picometres and 3.4 nanometres. The smaller periodicity, they proposed, represented the spacing of the bases, which were arranged at right angles to the sugar–phosphate chain. The larger spacing was generated because the chain was twisted in a helix with 3.4 nanometres between turns. Finally, they deduced that the long DNA molecule must contain more than one chain running parallel, and they therefore proposed that these chains must be intertwined.

In 1953, J. D. Watson and F. H. C. Crick put forward their model for the structure of DNA. This model proposed a structure which had the merit of being compatible with all the chemical and physical facts which had been deduced, while allowing the molecule to fulfil its two functions; information storage and the ability to replicate. Watson and Crick proposed that the DNA molecule consisted of two poly-nucleotide chains wound around one another, with the bases pointing to the axis of the helix. The spacing of the two chains was such that it imposed restrictions on which bases in one chain could lie opposite a particular base in the other chain. They proposed that adenine could only lie opposite thymine, or guanine lie opposite cytosine. These bases in the two chains were hydrogen bonded to one another. They also proposed that the two chains were of opposite polarity, such that if one had its deoxyribose residues linked $3' \rightarrow 5'$ by the phosphate groups, the other had them linked $5' \rightarrow 3'$. Because of the restrictions imposed by the specific **base pairing,** the base sequence in the two chains would be related, being **complementary** to each other. Watson and Crick pointed out that their structure not only allowed information to be stored in the sequence of **base pairs** in the chain, but also allowed for replication. If the two chains were to unwind from one another and each was to act as a template for the synthesis of a new chain, because of the base-pairing restrictions which have been described, two new molecules would be produced which were exact replicas of one another. Details of the hydrogen bonding between bases is given in figure 35, and the essential features of Watson and Crick's model for DNA structure are represented in figure 36. An important prediction which arises out of this, is that replication of DNA should be semi-conservative, that is a DNA molecule should have one of its nucleotide chains derived from its 'parent' molecule, but the other should be newly synthesised.

This prediction was tested in an extremely elegant way by M. Mesel-son and F. W. Stahl. These workers who published their results in 1958 grew the bacterium, *Escherichia coli*, for several generations on a medium in which the nitrogen source contained the heavy isotope of nitrogen, ^{15}N, instead of the usual ^{14}N. In this way all the nitrogen containing compounds in the cell, including the DNA, came to contain the ^{15}N isotope. They then transferred the bacterial cells to medium containing only the normal ^{14}N isotope. From time to time they withdrew samples from the culture, and extracted from them the DNA. They compared these various DNA samples using the tech-nique of caesium chloride density gradient centrifugation. This technique is a very sensitive one for separating molecules of different densities. Using it, it is possible to separate DNA containing ^{15}N, from DNA containing ^{14}N. Meselson and Stahl discovered that when the bacteria had undergone one cell division since the transfer to ^{14}N medium, the DNA had a density intermediate between that containing ^{15}N, and that containing ^{14}N only. This is consistent with

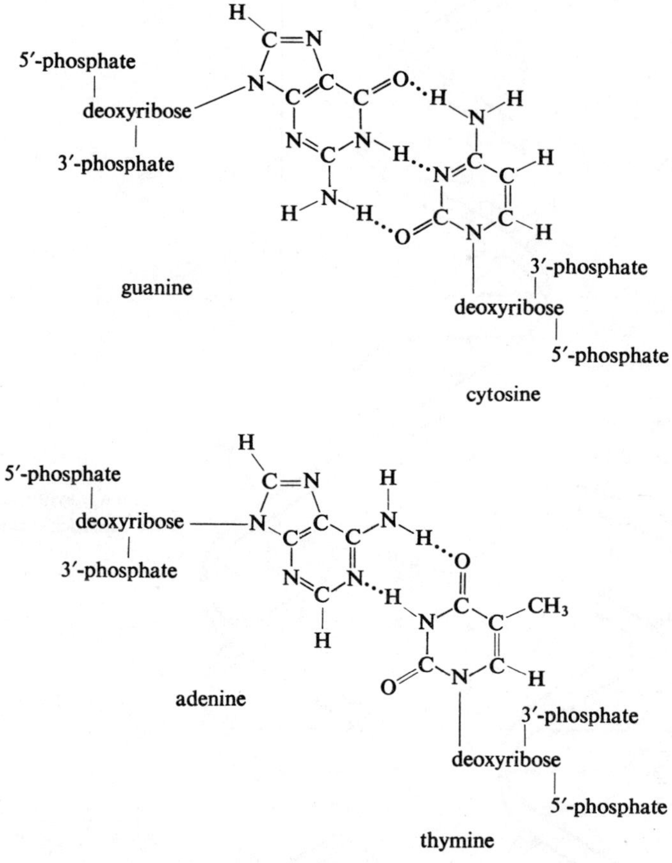

Figure 35. Hydrogen bonding between bases in DNA.

semi-conservative replication, as one chain of each DNA molecule would have been derived from the DNA in the bacteria at the time of transfer and would contain ^{15}N, and the other chain would have been made in the new medium and would contain ^{14}N. After the next cell division had taken place the DNA was of two densities, half had the intermediate density, and half had the density characteristic of pure

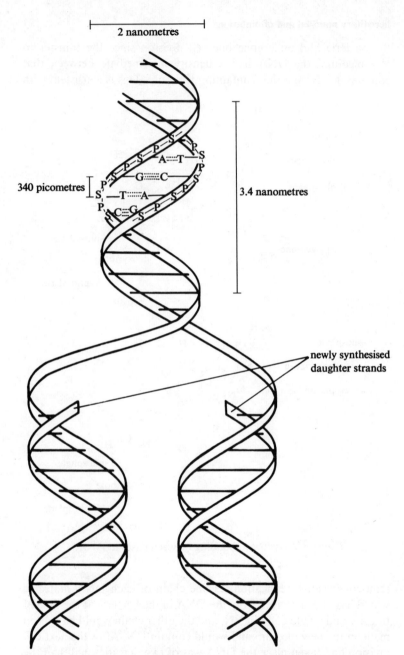

Figure 36. The structure of DNA proposed by J. D. Watson and F. H. C. Crick. The DNA molecule is shown replicating: the two strands separate and each directs the synthesis of a complementary daughter strand.

¹⁴N-containing DNA. This is again entirely consistent with semi-conservative replication. With successive cell generations the DNA was found to contain proportionally less of the intermediate density DNA and more of the ¹⁴N-containing DNA. These results are represented diagrammatically in figure 37.

Although Watson and Crick's model for the structure of DNA poses some problems, a great deal of evidence in its favour has been accumulated, and no serious challenge has been made to it. The precise molecular mechanism of replication, which on Watson and Crick's model, requires the unwinding of the DNA molecule at a rather high

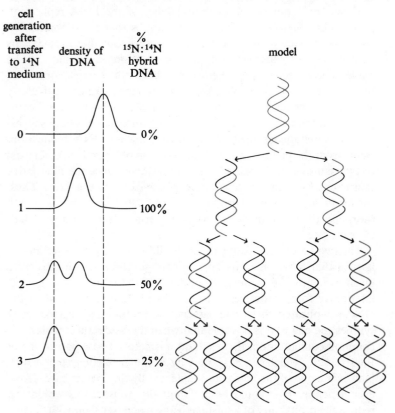

Figure 37. M. Meselson and F. W. Stahl's experiment to demonstrate the semi-conservative replication of DNA. A model for DNA replication is given on the right. The red strands represent ¹⁵N material, and the black, ¹⁴N material.

rate, is still not understood but it is thought that this process will not impose any requirements which are unacceptable on either chemical or physical grounds.

Having considered the structure and replication of DNA it is now time to turn to the mutation process. Clearly if it is to be proposed that the genetic information of the cell is stored in the base sequence of the DNA it contains, then it is logical to propose that mutations represent changes in this base sequence. We can now consider how these might occur. It has already been mentioned that mutations occur spontaneously at a very low frequency. This is taken as evidence in favour of the extreme fidelity of the DNA replication process, but there are now some indications that DNA replication itself may not be quite so accurate a process, but that mechanisms exist in the cell which are able to correct mistakes, should they be made. The cause of spontaneous mutation still remains something of a mystery. It is probable that there are many causes particularly as many of the agents known to induce mutations are natural components of the environment. One possible cause that has been suggested for spontaneous mistakes occurring during DNA replication, stems from the fact that each of the bases involved in DNA can exist in two tautomeric forms. This is important because the alternative tautomeric form pairs with a different base from the usual one. Thus, for example, adenine, whose normal tautomer hydrogen bonds with (normal) thymine, in its alternative form bonds with (normal) cytosine (see figure 38).

The two tautomeric forms are in equilibrium and this equilibrium is such that a base is in its normal configuration (i.e. the structure given in figure 33) for very much more of the time than it is in its alternative form. However if it is in its alternative form when the DNA is replicated, the wrong base will be incorporated into the newly synthesized chain. It is possible that when the base returns to its more common configuration, the resultant mismatched base pair is one of the structures which can be spotted by a cellular repair system, which may excise one of the bases, and replace it by the correct base. However if neither base could be preferentially excised, this would not reduce the probability of a mistake being made (see figure 39).

The study of factors which will induce mutation was initiated by H. J. Muller, who in 1927, demonstrated that X-rays could induce mutations in *Drosophila*. It was later shown that alpha, beta, and gamma rays were also mutagenic. All are ionising radiations, and have many properties in common. It is not surprising that these high-energy radiations, which are known to break chemical bonds should damage the DNA and cause mutation. These radiations are also known to cause extensive chromosomal damage, which can be seen using appropriate cytological techniques.

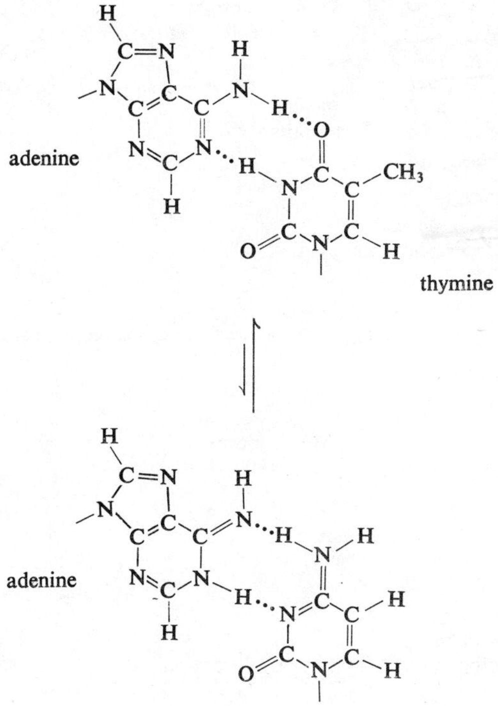

Figure 38. Hydrogen bonding of the tautomers of adenine.

Until now, we have been considering mutation as a process involving only a limited part of the base sequence of the DNA of one gene. Such mutations are referred to as **point mutations**. Mutation can also be the result of more extensive damage. Sections of the genetic material may be lost, in which case the mutation is called a **deletion**. Deletions may involve the loss of a part of a gene, or extend to several adjacent genes, such that the loss of chromosomal material is, in some tissues, observable. Mutation can also involve the reversal of a chromosomal segment within the whole chromosome, so that the gene order of a linkage group which could be represented *a-b-c-d-e-f-g-h*, might become *a-b-c-f-e-d-g-h*. This type of change is known as an **inversion**. Finally, a **translocation** may occur where a section of one

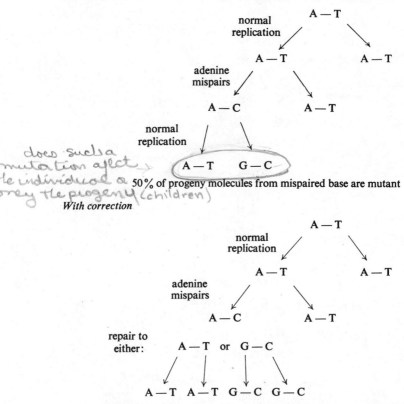

Without correction

normal replication

A — T

A — T   A — T

adenine mispairs

A — C   A — T

normal replication

A — T   G — C

50% of progeny molecules from mispaired base are mutant

does such a mutation affect the individual or only the progeny (children)

With correction

A — T

normal replication

A — T   A — T

adenine mispairs

A — C   A — T

repair to either: A — T   or   G — C

A — T   A — T   G — C   G — C

Provided alternative repair possibilities are equally frequent, 50% of progeny molecules from mispaired bases are mutant.

Figure 39. The consequences of base mispairing during DNA replication.

chromosome may be broken off and become attached to another chromosome.

Ionising radiations, as well as causing point mutations, commonly produce deletions, inversions and translocations. The details of the mechanisms involved are not however fully understood. A feature of the mutagenic activity of ionising radiations, is that in many systems it does not seem important whether a given dose of radiation is given at high intensity over a short period or at low intensity over a long period, each treatment causing the same frequency of mutation. This is of some importance, as the absence of a threshold means that the low levels of radiation, arising from radioactive fallout and other sources, to which everything is constantly exposed, will be as likely to cause a mutation as an equivalent intense dose given over a short period.

The mutagenic effect of ultraviolet irradiation is a little better understood. Whereas it is not even known whether ionising radiations have their effect by directly affecting the DNA, or by some indirect means, such as for example causing the production of some highly reactive short lived compound which can react with the DNA, it is thought that ultraviolet irradiation directly affects the DNA. It seems that, as a result of irradiation, the pyrimidine bases and in particular thymine, become altered so that there is a tendency for adjacent pyrimidine bases to cross link. The most common products are **thymine dimers.** It is known that there are repair mechanisms present in many cells which can excise these thymine dimers and repair the DNA. How exactly mutation is caused by ultraviolet light is still a matter of controversy. Possibly DNA containing thymine dimers cannot be replicated faithfully. Alternatively it might be that the repair system was not perfect, and mistakes were made during this process. Other different mechanisms might also be operative, as the mutagenic effects of ultraviolet irradiation are not confined to point mutation. In *Aspergillus* for example large doses of ultraviolet irradiation commonly cause translocations.

As well as radiations of various types, certain chemicals can also cause mutation. Chemical mutagens may be divided into two main classes. Firstly there are those which react with the DNA, altering it, so that subsequently mistakes are made when it is replicated. Such mutagens cause mutation not only when they are used to treat living cells, but also when transforming DNA, or free virus particles are treated. The second class of chemical mutagen is only effective at DNA replication. In this class are the base analogues which are incorporated into DNA at replication, but which do not have such constant pairing relationships as the normal bases.

The effects of some of the more common chemical mutagens will be described, and some consideration as to their possible mode of action will be given.

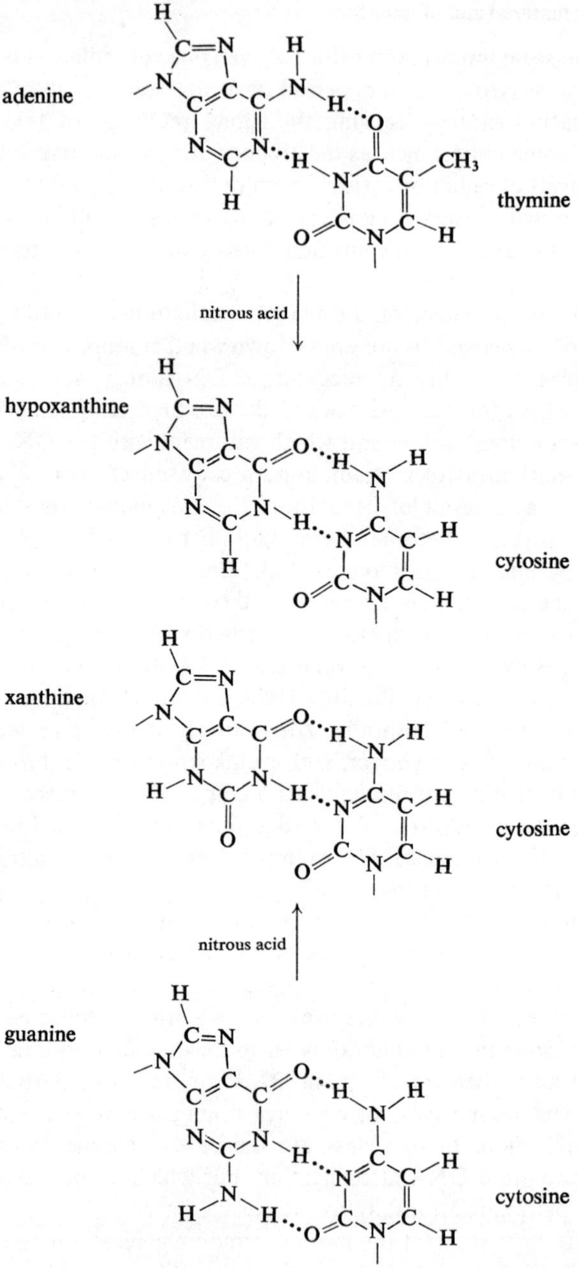

Figure 40. The mutagenic action of nitrous acid. Nitrous acid converts adenine to hypoxanthine and guanine to xanthine, both of which hydrogen bond with cytosine.

Chemicals reacting with DNA

(1) *Nitrous acid* is known to react with the bases in DNA and to deaminate them. The consequences of this are as follows. Adenine is deaminated to give hypoxanthine (see figure 40). At the next DNA replication, hypoxanthine pairs as though it was guanine, that is with cytosine, and hence an adenine–thymine base pair will be replaced by a guanine–cytosine. Guanine is deaminated by nitrous acid to give xanthine which also pairs as though it were guanine, thus there will be no mutagenic effect on guanine–cytosine base pairs. However nitrous acid deaminates cytosine to produce uracil (see figure 41). Uracil pairs with adenine, and so here a cytosine–guanine base pair will be changed to a thymine–adenine. Finally, nitrous acid cannot react with thymine, as it has no amino group.

The conversion of a purine to a purine, or a pyrimidine to a pyrimidine is often referred to as a **transition** whereas the conversion of a

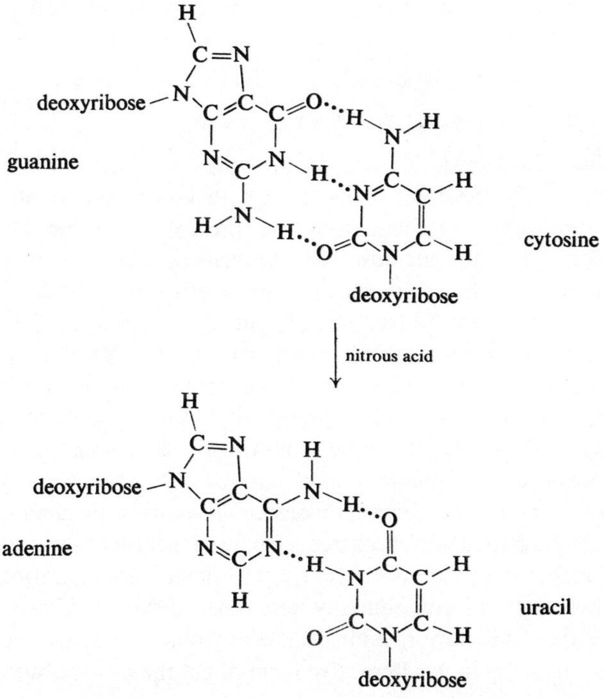

Figure 41. The mutagenic action of nitrous acid. Nitrous acid converts cytosine to uracil which hydrogen bonds with adenine.

purine to a pyrimidine or *vice versa* is called a **transversion**. Nitrous acid can only therefore induce transitions.

(2) *Hydroxylamine* appears to react with only one of the four bases in DNA, cytosine. There is some controversy as to what the product must be, but it is thought that this must pair with adenine at replication. Hydroxylamine will therefore result in the substitution of a thymine–adenine pair for a cytosine–guanine.

(3) *Alkylating agents*. This is a large class of mutagenic chemicals which include the nitrogen mustards, ethylene oxide, diethylsulphate, and nitroso-guanidine. These all have in common that they react with guanine, alkylating the 7 position. Diethyl sulphate produced 7-ethyl guanine for example. The 7-alkyl compound appears to have a labile linkage to the deoxyribose residue, and so tends to be released from the DNA. Such damage again might be repaired, but if such a process was prone to mistakes then any base might be substituted for the original guanine. This is consistent with the finding that alkylating agents cause both transitions and transversions, but only those involving guanine.

Chemicals mutagenic only at DNA replication

(1) *Base analogues*. Various compounds closely resembling the structure of the bases normally present in DNA, are found to be highly mutagenic, provided they are present when the DNA is replicated. They are not however mutagenic at other times, and so have no effect on free virus particles or transforming DNA. Two of the most potent are 5-bromouracil, and 2-aminopurine. 5-bromouracil, like the normal DNA bases, can exist in two tautomeric forms. Its more common form resembles thymine in its base pairing relationships, and can be incorporated instead of thymine into newly synthesised DNA. Under some conditions it is in fact possible to substitute thymine almost completely by 5-bromouracil. In its alternative tautomeric form, 5-bromouracil pairs with guanine (see figure 42). Its mutagenic effect stems from the fact that 5-bromouracil spends rather longer in its alternative tautomeric configuration than does thymine, and consequently generates mistakes in two ways. It may either be incorporated instead of thymine, but at a subsequent replication, being in its alternative form direct the incorporation of a guanine, or instead it may be incorporated in its rarer form instead of cytosine, but thereafter generally behave as though it were thymine.

In the former case the result would be the substitution of an adenine–
thymine base pair by a guanine–cytosine, and in the latter the substitution of a guanine–cytosine base pair by an adenine–thymine (see
figure 43). 5-bromouracil therefore causes transitions, but not transversions.

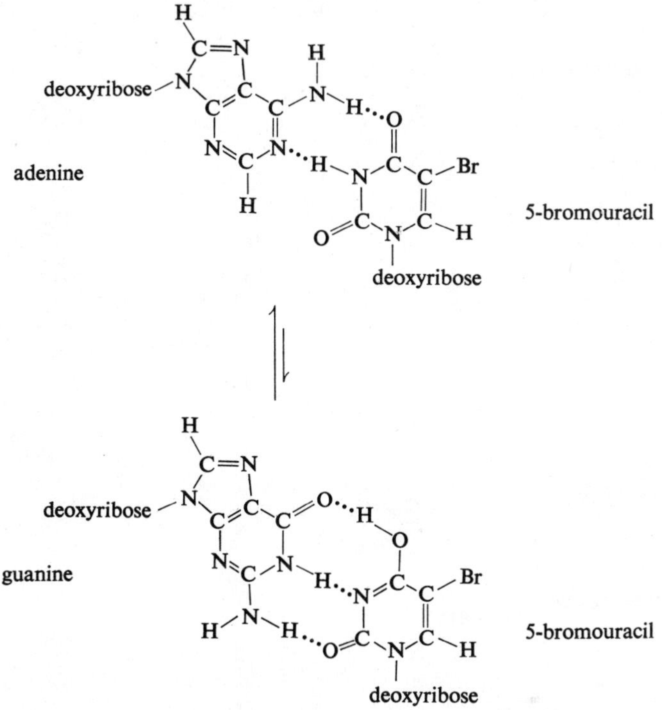

Figure 42. Hydrogen bonding of the tautomers of 5-bromouracil.

The mechanism of 2-aminopurine mutagenesis is thought to be
similar. In its normal form, 2-aminopurine pairs with thymine, but in
its alternative tautomeric configuration it pairs with cytosine. It can
also bond weakly to cytosine even in its normal state, and this is
probably one of the reasons why it is so effective a mutagen.

(2) *Acridine dyes.* A second important class of chemicals, the
acridine dyes, which include proflavin, are rather restricted in their
mutagenic effects. They appear to be appreciably mutagenic only to
bacteriophage which is multiplying within its host. Although some
theories have been put forward, it is far from clear why this should be.

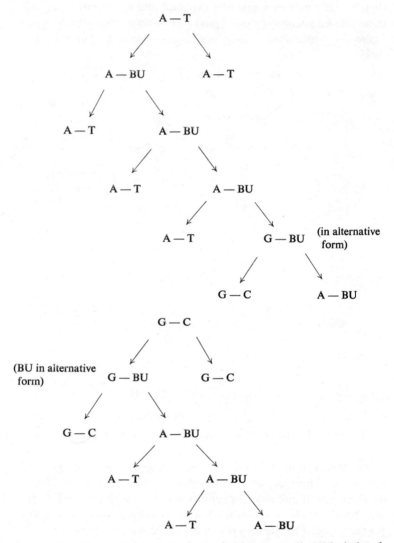

Figure 43. The possible mechanism of 5-bromouracil (BU) induced mutation.

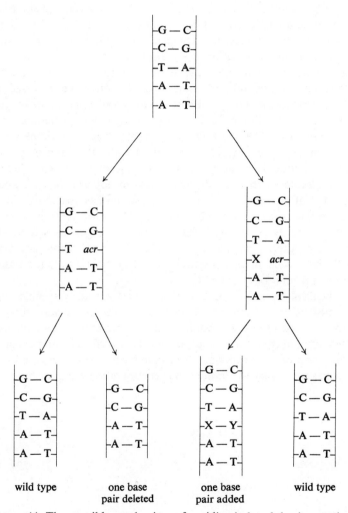

Figure 44. The possible mechanism of acridine-induced (*acr*) mutation.

Acridine dyes seem to act in a fundamentally different manner from base analogues or chemicals which react with the DNA. Whereas these produced base substitutions, acridines probably delete or add single bases to the DNA. Evidence for this is indirect, and some will be given in a later chapter. One significant point is that whereas it is often possible to reverse the mutagenic effect of, for example, a base analogue-induced mutation using another base analogue mutagen, base analogues will not reverse an acridine-induced mutation. However acridine-induced mutations can often be reversed by further treatment with acridines. It therefore appears likely that acridines do not induce base substitutions but small deletions of or additions to the DNA. Recently a possible basis for acridine action has been suggested. Acridine dyes react with DNA and may force their way into the molecule, between adjacent bases. This distorts the molecule so that the distance between adjacent bases is about doubled. If at replication, the template strand has associated with it an acridine molecule, this will cause an additional base to be inserted into the newly synthesised chain opposite the gap caused by this molecule. Alternatively, an acridine molecule may become associated with the new chain as it is synthesised, distorting it so that a base is omitted from it (see figure 44).

Chemical mutagens, because their effects are fairly specific, have enabled us to find out a great deal about the precise way in which DNA is used by the cell. Some of the experiments which have used chemical mutagens for this purpose will be described in later chapters. Since most of these experiments involve the use of viruses or bacteria it is next necessary to consider the genetics of these organisms.

6. The genetics of bacteria

The bacterial cell shows many structural differences from the eukaryotic cell. It is not surprising therefore that the genetic systems found in bacteria have no counterpart in eukaryotes. However sexual reproduction in both brings about the same end, the pooling of the genes from two different individuals, and their distribution in different combinations to their progeny. For a long time it was not thought that bacteria reproduced sexually. The process of transformation, described in the last chapter (see page 64) certainly caused recombinant progeny to be produced, but it was thought of as a laboratory curiosity rather than a natural method of bringing about recombination.

The first detection in bacteria of a more conventional type of sexual process was by J. Lederberg and E. L. Tatum, who published their findings in 1946. These workers set about searching systematically for a process which could give rise to recombinant progeny. They used the bacterium, *Escherichia coli*, upon which an enormous amount of biochemical research had already been done, arguing that this would be an invaluable foundation for their genetic studies. Lederberg and Tatum thought that bacteria might undergo sexual reproduction in the same way as a unicellular haploid eukaryotic organism does, two cells fusing to form a zygote, which then underwent meiosis. However they reasoned that if such a process existed it must be extremely uncommon, otherwise it would have been detected previously. They therefore devised a selective technique to detect rare progeny with recombinant genotypes. This technique required bacterial stocks which carried several mutations. Since these could not be synthesised by breeding experiments, they utilised sequential mutagenic treatments. They first treated a wild type with a mutagen, and obtained from it a strain carrying a biochemical mutation leading to a requirement for threonine. This threonine-requiring strain was then grown and treated again with mutagen, and a strain requiring both threonine and leucine was obtained from it. Finally this strain, after it had been multiplied, was treated once again with mutagen, and a strain having three requirements, for threonine, leucine and thiamin, was obtained. Using a similar procedure they isolated a strain requiring biotin, phenylalanine and cysteine. It was necessary to employ this step-wise procedure, because mutation in one gene occurs independently of mutation in others. Thus if mutations leading to a particular bio-chemical requirement occur in wild-type bacteria, which have been treated with a mutagen, at a frequency of 1 in 10^4 cells, the chance of obtaining a triply-requiring mutant strain after one treatment will be only 1 in 10^{12}, which is too low a frequency to be detected readily. By adopting a step-wise approach, the frequency of obtaining the required strain at each step was 1 in 10^4, which was a practical one for detection.

Lederberg and Tatum mixed these two triply-requiring strains together, grew them for a short period, and then plated them onto a medium which contained none of the six biochemicals which the original strains required. They found that about 1 in every 10^6–10^7 bacteria which were plated onto this selective medium could grow. On subculture these non-requiring strains retained their ability to grow on unsupplemented medium. They argued that these strains must have arisen by some recombination process. The use of triply-mutant parent strains precluded the origin of non-requiring strains by mutation. Each of the biochemical mutants could undergo reverse mutation (could **revert**) to give non-requiring strains and as with forward mutation, this reverse mutation occurs independently at each gene. In the absence of a mutagenic treatment the rate of re-version for each mutant gene is about 1 in 10^6 cells so that the rate at which a triply-mutant strain will revert to a non-requiring strain is only 1 in 10^{18} cells. Thus reverse mutation could not have accounted for the rate at which non-requiring progeny occurred in Lederberg and Tatum's experiment.

After this initial experiment, which was repeated using different combinations of mutations, further progress at elucidating the genetic system involved was slow. Among the mutations used were some which lead to drug resistance, and it was experiments using these which provided the next clue about the mechanism involved. It was found that if a streptomycin-resistant strain was derived from a stock A, and this was mixed with stock B in the presence of strepto-mycin, then recombinant progeny were obtained at the usual rate for stock A by stock B crosses. If however a streptomycin-resistant strain was derived from stock B, and mixed with stock A in the presence of streptomycin, no recombinants were obtained. It therefore appears that the continued growth of stock A was necessary in order to obtain recombinants, whilst that of stock B was not necessary, as recombinants were only obtained when stock A was protected from the streptomycin by a resistance mutation. From this grew the idea that in a bacterial cross the two parents were not equivalent. One was thought of as donating genetic information to the other, which had to be capable of continued growth in order to give rise to recombinants. In the example above, stock B would be the donor strain, and stock A the recipient. This is a somewhat similar situation to that found in transformation; however it is in this case necessary to have cell-to-cell contact in order to obtain recombinants.

The next step in the elucidation of this mating system, which was given the name **conjugation**, was the finding that donor strains could convert recipient strains to donor, without the transfer of other genes. Because of this, it was thought that the ability to donate genes must be due to some transmissible agent, which was called a fertility agent, or **F agent**. Donor strains were designated F^+, and recipient strains F^-. Only $F^+ \times F^-$ crosses were fertile and able to give rise to recombinants. When F^+ and F^- strains were mixed it could be seen by microscopic examination that the cells of the two strains paired with one another. Next, it was found that another type of strain could be obtained from F^+ strains. This new type of strain, called **Hfr**, gave rise to recombinants at a very much higher frequency than F^+ when mated to F^- strains, in some crosses recombinants occurring at a thousand times the frequency. Hfr strains, like F^+ strains, were found to pair with F^- strains. A detailed analysis of Hfr $\times$ F^- crosses revealed a complex pattern, with little order apparent. Some crosses yielded markedly more recombinants than others, but it appeared that all genes were part of a single linkage group. There was much controversy about the nature of the mechanism which could be giving rise to such results but it was eventually elucidated by F. Jacob and E. L. Wollman who devised a new technique, the **interrupted mating experiment**. This involved mixing cultures of an F^- and Hfr strain, and then at intervals after mixing, withdrawing samples, and subjecting these, after dilution, to vigorous mechanical agitation. The agitation has the effect of breaking apart paired bacterial cells, and the dilution ensures that their concentration is reduced to a level where it is unlikely that the pairs will reform. The diluted, disrupted samples were then transferred to medium which could be used to detect whether the Hfr had been able to donate its genes to the F⁻ before the mating process was interrupted. It is perhaps easiest to describe how this was done by using a specific example.

In a typical interrupted mating experiment the genotypes of the Hfr and F⁻ strains used might be as follows:

> *Hfr H* Wild type (i.e. able to grow without biochemical
>   supplements, but sensitive to streptomycin)
> F^- *strep^R thr   leu   pro   gal*

where *strep^R* represents a mutation conferring streptomycin resistance, *thr*, *leu* and *pro* are mutations leading respectively to requirements for threonine, leucine and proline, and *gal* is a mutation leading to the inability to utilise the sugar galactose as a carbon source. These two strains are mixed, and samples withdrawn at 5-minute intervals. After dilution and agitation, these are plated on a series of media to test for the transfer from the Hfr to the F⁻ of the wild-type allele of the various genes involved. To test, for example, for transfer of the $+^{thr}$ allele, the sample would be plated on medium containing leucine and proline, having glucose as a carbon source, and containing streptomycin. Since the medium contains streptomycin, the Hfr strain will be unable to grow, and since it contains no threonine the F⁻ strain will be unable to grow either. Only recombinant strains which have received the wild-type allele of the threonine gene will be able to grow. The transfer of the other alleles can be tested using similar types of media. Thus $+^{gal}$ transfer can be assayed by plating on medium containing threonine, leucine, proline and streptomycin, in which the carbon source is galactose. As before only recombinants will grow, but here they will have had to receive the $+^{gal}$ allele from the Hfr.

The results obtained for such experiments may be represented graphically by plotting the number of recombinants receiving a particular allele from the Hfr for each time of sampling. When this is done characteristic graphs are obtained of the type shown in figure 45. It will be seen that the curve obtained for each gene is basically the same. First there is a period during which no recombinants which have received the gene can be detected. This is then followed by a rapid rise to a plateau value, which is then maintained. It can also be seen that the longer the lag before a recombinant is detected, the slower is the rise to the plateau, and the lower is the plateau value obtained.

To explain results such as these, Jacob and Wollman proposed that the transfer of genes from the Hfr to the F⁻ must proceed in a fixed order, and must be a fairly lengthy process. Upon mixing, the F⁻ and Hfr strains would pair. The Hfr would then begin to transfer its

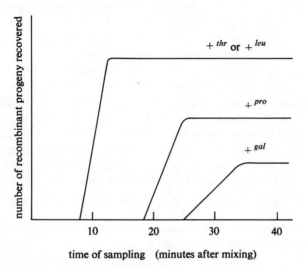

Figure 45. Results of a typical interrupted mating experiment. (For details see text.)

genes to the F⁻. With the example set out above, it must take 8 minutes before the Hfr can transfer a $+^{thr}$ or $+^{leu}$ gene to the F⁻, for if the mating is interrupted before that time no recombinants for these genes are obtained. Similarly the $+^{pro}$ gene cannot be transferred until 18 minutes have elapsed, and the $+^{gal}$ gene until 25 minutes have passed. The genes can therefore be ordered linearly on a time basis as shown below.

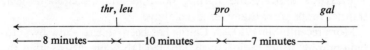

This time map was consistent with the linkage map obtained using the pooled recombination data from previous crosses. Jacob and Wollman pointed out that the rise to the plateau level would not be instantaneous unless all the bacteria paired at the same moment, and the Hfrs transferred their genes at exactly the same rate. The slope of the rise was a reflection of the variability within the population and in particular, the slight variations in the rate of gene transfer would account for these slopes becoming progressively less steep as genes with a later time of entry were observed. Finally they argued that the plateau values obtained were an indication of the final number of mating pairs which successfully transferred a particular gene. These plateau values were lower for late entry genes, because mating pairs could separate spontaneously during the mating process thereby reducing the total number of pairs still able to continue gene transfer.

Two other important observations were made using interrupted mating experiments. Firstly it was found that Hfrs of different origin did not all behave alike. Each Hfr donated its genes in a characteristic order, and although the order obtained with each Hfr was consistent with that obtained with another, it did not necessarily begin at the same place, nor were the genes necessarily donated in the same direction. One Hfr might for example donate its genes in the order $a\,b\,c\,d\,e\,f\,g\,h\,i$, another $f\,e\,d\,c\,b\,a\,i\,h\,g$, and a third $f\,g\,h\,i\,a\,b\,c\,d\,e$.

Each of these orders can be made consistent with the same map if this is circular as shown in the diagram.

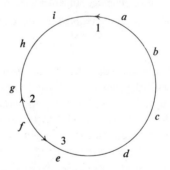

The arrows indicate the starting position and direction of gene transfer in the three examples given above.

More recently (in 1963), J. Cairns has been able to show by auto-radiography, using radioactively labelled DNA precursors, that the DNA in *E. coli* is in the form of large circular molecules. These molecules have a contour length of 1.1 mm (compared with the average size of an *E. coli* cell which is approximately $1.5 \times 0.5\ \mu\text{m}$). This size of DNA molecule is thought to be sufficiently large to store the total *E. coli* genetic information, and so it is obviously attractive to equate the circular map obtained by interrupted mating experiments with this circular DNA molecule.

The other point to emerge from interrupted mating experiments is that the majority of the F$^-$ bacteria which have received genes from an Hfr, remain F$^-$. Only if an Hfr succeeds in donating its entire genome, a process which at 37 °C under ideal conditions takes approximately 90 minutes, will the recipient F$^-$ be converted to an Hfr. From this emerged the idea, that the F agent must be responsible for the initiation of gene transfer, perhaps breaking the circular DNA molecule at a particular point which varies from one Hfr to another. However because the F$^-$ is only converted to an Hfr after total gene transfer, it must be concluded that at least some part of the F agent is not transferred until the very end of the conjugation process. There is evidence that the F agent itself consists of DNA and so it is possible to put forward an attractive and relatively simple model for the various facts about conjugation which have been collected. This is as follows:

(1) The hereditary information is contained in one large circular DNA molecule.

(2) F$^+$ strains contain as well as this DNA molecule, another smaller and also possibly circular DNA molecule, the F agent.

(3) The F agent is replicated independently of the larger molecule.

(4) The F agent carries information which makes strains containing it pair with strains which do not (F$^-$ strains).

(5) Once paired, the F agent is transferred from the F$^+$ to the F$^-$ through a small bridge connection which can be seen by electron microscopy. It is probable that this transfer is accompanied by F agent replication, as the F$^+$ remains F$^+$ after transfer. This process is represented in a highly diagrammatic way in figure 46.

(6) In an F$^+$, the F agent can pair with the main bacterial DNA molecule at a number of places, and then by a process of recombination become integrated to give a single circular hybrid molecule. The F$^+$ has now become an Hfr. This mode of origin of Hfrs is represented in figure 47.

(7) The integrated F agent in the Hfr still retains its properties of causing the cell to pair with an F$^-$ cell, and also of transferring itself to the F$^-$ by replication. However if the Hfr DNA molecule is regarded as an F agent, with the main bacterial DNA inserted in it, it will be seen that in order to transfer the whole of the F agent, all the

bacterial DNA will have to be transferred as well. Furthermore the recipient F^- will only receive the whole of the F agent if all of the Hfr's genetic information has been transferred. This transfer process is shown diagrammatically in figure 48.

(8) Finally in a similar manner to the way Hfrs can arise from F^+ by recombination, the reverse process can occur, and an F^+ can be produced from an Hfr by the elimination of the F agent from the bacterial DNA.

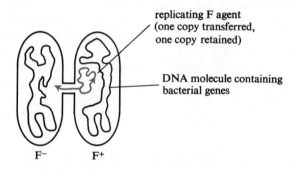

Figure 46. Transfer of F agent from F^+ to F^- strain.

This model is consistent with all the facts that have been outlined previously. It explains for example why $F^+ \times F^-$ crosses give rise to recombinants at a low but detectable rate, as an F^+ culture will always contain a very small proportion of Hfrs which have arisen by recombination. It will also be seen why it was so difficult to unravel the mechanism of bacterial conjugation, and why results from $F^+ \times F^-$ crosses were so variable.

It is possible to calculate the rate of DNA transfer during conjugation. Using the contour length obtained by autoradiography, it can be deduced that there are approximately 3 million base pairs in the bacterial DNA molecule. Since complete transfer takes 90 minutes, the rate of transfer must be about 35 000 base pairs per minute. Although this rate may seem high, there are no objections to it on physical or chemical grounds.

The consequences of conjugation have not yet been considered but it will be appreciated that the product of conjugation will not usually correspond to a zygote in a eukaryotic organism, as it will not contain two whole genomes. For this reason it is usually referred to as a **heterogenote** rather than a zygote. It appears that the fragment of

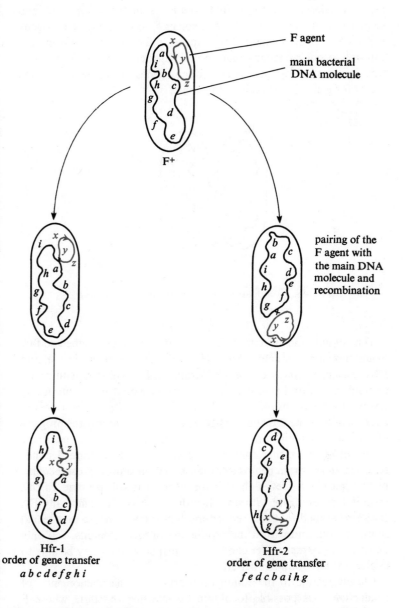

Figure 47. Mode of origin of two different Hfr strains from an F⁺. The letters *a, b*, etc. represent the location of specific genes.

DNA transferred from the Hfr to the F⁻ is unable to replicate, and so on subsequent division of the recipient F⁻ only one of the progeny will receive this fragment. Integration of the donated information can only occur by recombination, and furthermore two recombinational events will be necessary if donor genes are to be substituted stably for genes in the recipient genome.

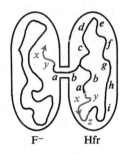

Figure 48. Conjugation between Hfr and F⁻ bacteria.

The conjugation system in *E. coli*, as it has been described so far, would have a major defect. Since Hfr × F⁺ crosses are not fertile, and F⁺ strains can rapidly convert F⁻ strains to F⁺, whilst remaining F⁺ themselves, natural populations of bacteria would not contain any fertile combinations of cells. This apparent paradox is resolved because of two phenomena. Firstly, when growing rapidly the progeny of some F⁺ cells fail to receive an F agent, and so are F⁻, and a new supply of F⁻ cells is therefore always present in an F⁺ culture. Secondly, when an F⁺ strain begins to exhaust its food supply, it can become phenotypically F⁻, even though the F agent is still present. In this state it can conjugate with an Hfr although it is still genetically F⁺. It can be argued that this second mechanism will ensure that sexual reproduction, and with it the production of recombinants, will only occur to any extent under conditions when new types may be able to exploit the environment better.

Closely related to the conjugation process is the phenomenon of **sexduction**. It is possible to obtain by selection, bacteria whose F agents contain a few genes only of the bacterial genome. Such F agents are designated F′. F′ agents are obtained by interrupting mating after a short period, and selecting for the transfer of a gene, which is not normally donated in conjugation until very late. When this is done, bacteria are obtained which are genetically unstable, but which are able to transfer the selected genes to F^- bacteria with the same efficiency as normal F^+ bacteria can transfer the F agent. It is this process which is called sexduction. The basis of sexduction is thought to be that F' agents are produced as a result of mistakes in the recombination process by which Hfrs can give rise to F^+ bacteria. If this reverse recombination event does not occur at the same spot as the forward event by which the Hfr was formed, an F agent containing some bacterial genes will be produced (see figure 49).

As the F agent is capable of division, an F^- bacteria which has received an F' agent will be diploid for a small portion of the bacterial genome, and will remain so upon division. These partial diploids are one of the ways in which complementation and dominance tests can be carried out in bacteria, and have been exploited in several important pieces of work connected with the study of how and when genes work. Some of these experiments will be described in detail later in this book.

The process of conjugation as described above has been elucidated almost entirely as a result of experiments on the bacterium *Escherichia coli*. Similar processes, which have been studied less extensively, have been detected in a number of species of bacteria, of which most are generally held to be fairly closely related to *E. coli*. There remain a large number of species still to be studied, and it is possible that other systems of bacterial gene transfer quite distinct from conjugation have yet to be discovered.

In fact one further method of gene transfer of an entirely novel type has already been elucidated. This is the process of **transduction**, in which a virus plays an integral role. Transduction was first detected in the bacteria *Salmonella typhimurium* by N. D. Zinder and J. Lederberg, who reported their finding in 1952. These workers found that it was possible to obtain wild-type recombinations from mixtures of two strains carrying biochemical mutations in just the same way as it was in *E. coli*. However they discovered a major difference, as cell-to-cell contact between the strains was not necessary. Although in this respect there appeared to be similarities to transformation, it was found that in this case there was evidence of polarity, as only the cell-free extract of one strain could bring about genetic alterations in the other, and not *vice versa*. An attempt was made to purify the

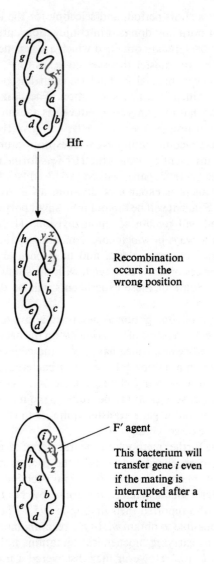

Figure 49. Mode of origin of F′ agents.

agent responsible, and it was found to be inseparable from a bacterio-phage, P22.

A more detailed analysis showed that for genes to be transferred it was necessary for the 'phage to inject its DNA into the recipient, but it need not necessarily be able to multiply there. This suggested that the 'phage particle was acting as a vector, carrying bacterial genes as well as its own. It was found that the probability of transfer of a number of different bacterial genes was about the same, but each was transferred independently. However it was later discovered that some genes could be transduced together (**co-transduced**). It is logical to regard such genes as being linked, so that the bacterial DNA carried by the vector 'phage, contains the two genes. Using co-transduction frequencies, linkage maps can be established. Trans-duction has proved to be particularly useful for the detailed genetic mapping of small regions of the bacterial genome, but, because the transducing 'phage can only carry a rather small portion (about 1 %) of the bacterial genome, it is not an ideal technique for establishing the latter's gross structure.

The mechanism of transduction is difficult to discuss without first considering certain aspects of the physiology of bacteriophage infection, which will be done in the next chapter. However it is in many respects thought to be analogous to sexduction, where it will be remembered the F agent can become the vector of bacteria genes (see page 94). As will be seen in the next chapter, some 'phage are able to form stable relationships with bacteria, possibly by the recombination of the two genomes. Subsequently this process can be reversed, and it is likely that this sometimes occurs inaccurately with the result that 'phage particles are produced which carry some bacterial genes (cf. F' agents). Some types of 'phage are only able to attach themselves to the bacterial genome at one or a few locations. Such 'phage can only transduce those bacterial genes near to their attachment site, that is they are said to show **restricted transduction**. Some other 'phage types however appear to be able to transduce any bacterial gene. These 'phage are said to show **generalised transduction**, and it is possible that this is only because they have many more alternative attachment sites. However it is also possible that a rather different mechanism might be involved in generalised transduction from that in restricted transduction.

As with the other methods of bacterial gene transfer described, the product of transduction will not be a complete zygote, but a heterogenote. As before, the integration of the transduced genes will require recombination. Transduction has only been characterised in a small number of bacterial species which include *Escherichia coli*. It is possible of course that transduction-like processes may be encountered with some eukaryote virus systems although none has yet been reported. This may be because in these systems the virus is unable to enter into the same sort of stable relationship with the host, but the distinct possibility of the existence of such phenomena must still remain.

F agents and viruses able to bring about transduction have certain similarities, both being able to associate themselves with the bacterial DNA, or instead to replicate independently in the cytoplasm. Such genetic elements are called **episomes**. There are a number of other such elements which are related to episomes, although there is no conclusive proof that these become integrated stably into the bacterial DNA. Some of these contain genetic information which not only directs the synthesis of compounds toxic to bacteria, but also confers resistance to the toxin on the bacterial cell in which the element is contained. Another type of genetic element, the **resistance transfer factor** (RTF), is of considerable importance to man. RTFs contain genetic information which renders bacteria which contain them resistant to, in some cases, as many as eight different antibiotics. Since RTFs are transmissible in much the same way as F agents, they can spread rapidly through a population of bacteria, all the cells becoming resistant. RTFs can also be transmitted to bacteria of related species, and this constitutes a particular hazard to man, as RTFs present in non-pathogenic species such as *Escherichia coli* can be transmitted to such pathogens as *Salmonella typhi*, the cause of typhoid fever.

One final point about a term which has been avoided in this chapter. It will be found that a large number of people, including those actively carrying out research in the field of bacterial genetics, refer to the bacterial DNA as a chromosome. It should be remembered that the bacterial chromosome is a very much simpler structure than the eukaryotic chromosome, whose behaviour was described in chapter 3. Attempts have been made to introduce alternative terms for the bacterial chromosome, but none has become generally used. It will also be found that the viral genetic information, which is a still simpler structure, is sometimes referred to as a viral chromosome.

7. The genetics of viruses

Viruses, being very specialised parasites which are unable to grow outside the cells of their host, are often difficult to handle experimentally. Little is known therefore about the genetics of a large proportion of viruses. Some have been studied however and it is bacteriophage, the bacterial viruses, which are by far the most studied group. In fact it is the 'phages of *Escherichia coli* whose genetic systems are best characterised, a further example of the advantage of using a system where much of the ground-work had already been done. Recombination in bacteriophage was first reported in 1946, independently both by M. Delbrück and W. T. Bailey, and by A. D. Hershey. Before the characteristics of the 'phage recombination process, and the important discoveries it has led to, can be discussed it is necessary to consider certain aspects of the 'phage structure and physiology.

'Phage Morphology

Bacteriophages vary considerably in their size and shape, but the 'phages which will be concentrated on in this chapter are relatively large. Bacteriophage T2 is about 200 nm × 65 nm (compared with an *E. coli* cell which is very approximately $1.5 \times 0.5 \ \mu$m). Not all 'phages contain DNA, some containing RNA instead, but the 'phages which have been most studied genetically are the DNA 'phages. As an example of the anatomy of one of these larger DNA 'phages, T2 will be used. T2 has a head and tail (see figure 50). The head consists of a core of DNA and a protein coat containing one type of protein molecule only. The tail is a specialised device for the injection of DNA into the host cell. It has a hollow central core surrounded by a sheath, both protein. At the end away from the head is a base plate, and long tail fibres which may be wrapped round the tail. Both of these components are also protein.

Other similar 'phage show variations to this structure. 'Phage λ for example has a thinner tail, and does not appear to have a sheath around it, the base plate is less elaborate, and there are no tail fibres.

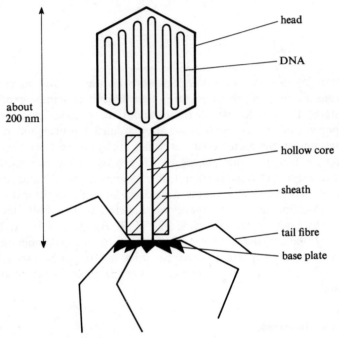

Figure 50. Diagram showing the morphology of 'phage T2.

Life cycle

A 'phage particle first attaches itself to the wall of a sensitive bacterium by its tail fibres. It is probable that this involves a fairly specific antigen–antibody type interaction, as it is possible to obtain mutant 'phage with altered tail fibres which show changes in their ability to adsorb to bacteria. It is also probable that the basis of some bacterial mutations to 'phage resistance lies in changes in the chemical composition of the bacterial cell wall.

Once attached, the virus particle releases an enzyme, **lysozyme**, which hydrolyses the bacterial cell wall, making a hole in it. Possibly triggered by the products of this hydrolysis, the 'phage's tail sheath then contracts, forcing the hollow core through into the bacterial cell. The DNA now passes into the host, and as was shown by the experiment of Hershey and Chase (see page 66) it appears that little if anything other than DNA passes into the bacterium. Once the virus DNA is inside the bacterial cell, a number of things can happen to it depending on the type of bacteriophage involved and the state of the host cell. These various fates will be dealt with in turn.

(1) **Virulent infection.** When infection is virulent, the 'phage DNA takes control of the bacterial metabolism. With 'phages such as T2, a number of 'phage enzymes are first made, then after 6 minutes (at 37 °C) synthesis of 'phage DNA begins. After about 12 minutes, new complete 'phage particles can be detected for the first time, but the bacterial cell is not normally **lysed** until about 30 minutes after infection, when it breaks open to release up to 300 new 'phage particles. The number of progeny 'phage produced from a single infection is referred to as the **burst size** and this varies with the type of 'phage used. The burst size for a particular virus can also be varied by altering the growth conditions during the infection process.

Virulent infection may be detected in two ways. The first is a simple qualitative way of testing for the presence of infective virulent 'phage, and consists merely of adding the sample to be tested to a dense culture of bacteria. After incubation for an hour or so the bacteria will be killed, and the culture will become clear. It is also possible to assay for the presence of virulent bacteriophage quantitatively. A few 'phage particles are mixed with a large excess of sensitive bacteria, and the mixture is spread on the surface of medium in a petri-dish. The bacteria will grow and cover the plate except for a few clear areas called **plaques**. Each of these is the result of the multiplication of a single 'phage particle in the original sample, its progeny infecting and killing a progressively larger area of bacteria, which shows up as the circular clear plaque.

(2) **Temperate infection.** Not all bacteriophage are capable of temperate infection. When the DNA of those that are enters the host cell, infection may be virulent, however sometimes instead the DNA becomes associated with the host DNA and replicates in step with it. The host cell is not destroyed, and is apparently unaffected by the presence of the 'phage. This integration process is called **lysogeny,** and bacteria carrying a bacteriophage in the integrated or **pro-phage** state are said to be **lysogenic** for that particular type of bacteriophage. The lysogenic state is a fairly stable one, the bacteria dividing, and all the progeny bacteria receiving a pro-phage. Every

so often however (about 1 in every 1000 cells), the pro-phage becomes dissociated from the bacterial DNA, and initiates a virulent infection cycle, so that the cell is killed and progeny 'phage released. For this reason cultures of lysogenic bacteria usually contain free 'phage particles. It is possible in the case of some bacteriophage, by various treatments which include ultraviolet irradiation, to cause the pro-phage in almost all the bacteria in a culture of a lysogenic strain, to dissociate and begin virulent infection, so that large numbers of bacteriophage particles are released. The contrast between virulent and temperate infection is shown in figure 51.

(3) The third possible fate of infecting 'phage DNA follows on from the temperate infection process. It is found that lysogenic bacteria are immune to further infection by 'phage of the same type as they carry in the integrated pro-phage state. The newly infected bacteriophage DNA is unable to divide, and will get diluted out by subsequent division of the bacteria.

Viruses capable of temperate infection can be recognised because, although they form plaques, these are not clear like those formed by virulent 'phage, but are turbid, because resistant lysogenic bacteria remain in the region of the plaque.

It is thought that the integration process involved in lysogeny may resemble F agent incorporation in Hfr formation. In some cases it can be shown that the bacteriophage DNA has recombined with the bacterial DNA to form a single continuous molecule.

As in other organisms, it is necessary to have recognisably different strains of bacteriophage before it is possible to investigate their genetic mechanisms. The range of characters which can be used in 'phage is rather limited. Two of the most important have been host range and plaque morphology.

It is possible to get mutant 'phage which differ from normal 'phage in the range of host strains they are able to infect. These can be shown by the use of techniques to be described below to carry mutations, all of which are located in the same gene. This gene appears to contain the information for the structure of the 'phage tail fibres, which confer the specificity of infection to the 'phage. The most common form of mutation affecting plaque morphology is the r mutation. Strains carrying r mutations, when plated with certain strains of bacteria form plaques which are larger and have more clearly defined edges than those produced by wild-type 'phage.

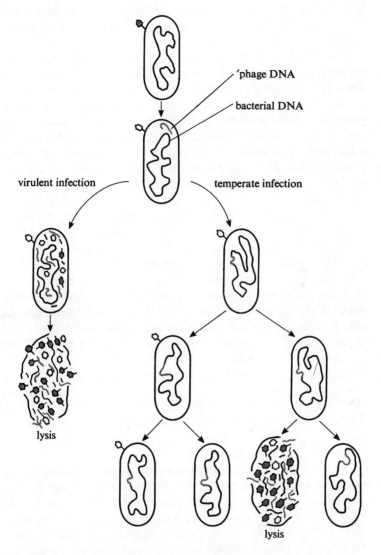

Figure 51. Virulent and temperate infection of bacteria by 'phage.

Mutation in three separate genes can lead to the *r* phenotype, and one of these, the *rII* gene has been especially important in the genetic analysis of 'phage.

As well as those mutations affecting host range and plaque morphology, it is possible to get various other types of 'phage mutant. These include mutants resistant to osmotic shock, which are found to produce an altered head protein and, perhaps the most useful, **conditional lethal mutants**. Temperature-sensitive mutants, which have been described previously (see page 45) represent one class of conditional lethal mutant. At the permissive low temperature growth is normal, but at a high temperature no growth can occur. It is possible to get temperature-sensitive mutations in a large number of 'phage genes, and it is not necessary to know their physiological basis to use them. Another type of conditional lethal mutant, the *amber* mutant is commonly used in the genetic analysis of 'phage. A description of this type of mutant can be found in a later chapter (see page 148), since it also occurs in other organisms.

'Phage crosses are very simple to carry out. The two parent strains of 'phage are added to a bacterial strain sensitive to both, at a sufficiently high concentration that it is likely that each bacterium will absorb several particles of each of the two 'phage strains. In the case of some types of 'phage, including the genetically important T2 and T4, the presence of 'phage within a bacterium normally prevents the successful entry of more 'phage. This is obviously a bar to performing a cross, but fortunately it is possible to eliminate this effect by carrying out the infection in media where bacterial metabolism is inhibited. A medium containing cyanide is usually employed for this purpose; bacterial metabolism, and with it 'phage development, being re-initiated by dilution of the cyanide.

The analysis of a 'phage cross is simply done by waiting for the culture of mixedly infected bacteria to lyse, collecting the progeny 'phage, and plating a suitable indicator bacteria to score for the various genotypes. When this is done it is found that recombinants occur with a characteristic frequency and that in two-factor crosses the two types of recombinant occur in equal numbers. In this way linkage maps may be built up. 'Phages T2 and T4 have a single circular linkage group. This need not indicate that the 'phage DNA is itself physically circular, and there is evidence that this is not so. It is found that in T2 and T4, the terminal parts of the linear DNA molecule are the same. This can be represented as follows:

abcdefghab.

It is thought that each 'phage contains a **circular permutation** of the same sequence. Thus two more 'phage particles might have DNA which can be represented:

fghabcdefg   cdefghabcd.

This circular permutation probably originates because the 'phage DNA is synthesised as a continuous end-to-end super polymer or **concatenate**, which is then chopped up to give the individual 'phage DNA molecules as follows.

↓   ↓   ↓   ↓   ↓
abcdefghabcdefghabcdefghabcdefghabcdefgha

↓
abcdefghab   cdefghabcd   efghabcdef   ghabcdefgh

When the physiological effect of a mutation is correlated with its linkage relationship, an interesting fact emerges. It appears that the linkage relationships are not random. For example all those genes which are necessary for some function early in 'phage development map together in one region. The significance of this grouping together of functionally related genes is still not completely understood; however a similar situation is found in bacteria, and the significance there will be discussed fully in a later chapter (see page 161).

Bacteriophage crosses differ from crosses in bacteria and eukaryotes in several important respects. In the account of virulent infection earlier in this chapter, it was stated that 'phage DNA synthesis begins 6 minutes after infection, complete 'phage particles can first be detected after 12 minutes, and that cell lysis does not usually occur until 30 minutes have elapsed. It appears that recombination can occur between 'phage DNA molecules once they are synthesised, and that molecules are available for recombination until incorporated into 'phage particles. Molecules which are the result of a recombinational event therefore have a chance of participating in a further recombination. That this is so can be shown rather simply; if a bacterial culture is mixedly infected with three different 'phage strains, which can be represented, $a\ b +$, $a + c$ and $+ b\ c$, it is possible to recover, after a single cycle of infection, 'phage with a $+ + +$ genotype. Such 'phage can only have originated by two successive recombinational events. As well as the possibility of recombinant genomes participating in further acts of recombination, they may also be replicated. This, and the random removal of DNA molecules into 'phage particles, together with variations which may occur in the course of virulent infection due to environmental effects, make the quantitative analysis of 'phage crosses difficult. Quantitatively 'phage genetics poses problems akin to those encountered in the study of the genetics of populations of eukaryotes. Despite these disadvantages, 'phage genetics has been of immense importance in the furtherance of our knowledge of the organisation of the genetic material at the sub-genic level. The potential of 'phage as a tool for the detailed analysis of genetic organisation was first exploited by S. Benzer in the 1950s.

Benzer studied mutations in the *rII* region of bacteriophage T4. These not only affect plaque morphology, but also change the host range of 'phage which carry them. An *rII* mutant produces large, sharp-edged plaques on *E. coli* strain *B*, but is unable to grow at all on *E. coli* strain K12, which is lysogenic for 'phage λ. Wild-type T4 is able to grow on both these strains giving small fuzzy-edged plaques on each (see figure 52). This straightway introduces one of the greatest advantages of this system. It is possible to select both for forward mutation, i.e. from $+\rightarrow rII$ and for reverse mutation, i.e. from $rII\rightarrow+$. Forward mutation can be detected by plating on *E. coli* strain *B*, where *rII* mutants are readily distinguishable from the wild type. Reverse mutation can be detected on *E. coli* K12 where only wild-type strains can grow.

Benzer showed that *rII* mutants were all closely linked, mapping in one region of the linkage map. He then carried out an analysis of the ability of *rII* mutants to complement one another. Complementation can be detected by mixedly infecting *E. coli* K12 with the two *rII* mutants to be tested. Since neither mutant on its own can infect

	E. coli strain	
	B	K12
wild type	small fuzzy-edged plaques	small fuzzy-edged plaques
rII	large sharp-edged plaques	no growth

Figure 52. Plaque morphology of wild type and *rII* mutant strains of 'phage T4 on different *Escherichia coli* strains.

and lyse this bacterial strain, lysis will only occur when the two 'phage strains are mutant in different genes, and are therefore able to complement one another. Using this test, Benzer found that all *rII* mutants fell into one of two groups, *A* or *B*, such that no complementation occurred within a group, but that all members of group *A* complemented all members of group *B*. Benzer called these groups **cistrons,** but they correspond to what have been referred to as genes throughout this book. It appears therefore that the *rII* region of 'phage T4 consists of two closely linked, complementary genes.

Benzer next went on to see whether recombination could occur between mutations which affected the same gene. This he was able to do because of the extremely high potential of genetic resolution which results from the ability to get progeny 'phage in vast numbers. His procedure for detecting rare recombinants among *rII* mutants was as follows.

Cells of *E. coli* strain *B* were first mixedly infected with two non-complementary strains of *rII* mutant. This mixture (appropriately diluted) was added to a dense culture of *E. coli* strain K12, and the whole plated out. If recombination was possible between the *rII* strains used, this would occur in the *E. coli* B strain where both virus strains could grow. These would then lyse to produce recombinant wild-type progeny which would be able to infect and lyse the *E. coli* strain K12, and as a result a plaque would be formed (see figure 53).

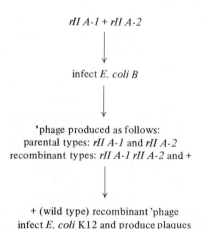

Figure 53. Recombination test for *rII* mutants of 'phage T4.

It is feasible to detect recombination frequencies as low as one recombinant per hundred million progeny using this technique. What Benzer in fact found was that recombination could occur among strains mutant in the same gene, but that if it occurred, the lowest rate of recombination under his standardised conditions was 1 in 10^4 or 0.01 %. Some mutants of different origins failed to give recombinants, even if 10^8 progeny were tested. Benzer argued that in the latter case the mutations involved must affect identical sites, and that the lowest recombination frequency he obtained, 0.01 %, must represent recombination between the basic mutable units in the DNA. It is possible to get an idea of the physical nature of what these units might be. The molecular weight of the T4 DNA molecule can be determined, and from this it can be estimated that the molecule consists of 150 000 base pairs. The total map length of the T4 linkage group is about 800 map units, where 1 map unit is the distance separating genes which show a recombination fraction of 1 %. The lowest recombination frequency, 0.01 %, represents 1/80 000 of the total map length, which is about two base pairs. Therefore it is probable that recombination can occur between base pairs one or two places apart in the DNA. More recently an elegant direct proof that recombination can indeed occur between adjacent base pairs in a DNA molecule has been provided by C. Yanofsky.

Benzer found that the recombination data was consistent with the two genes being linear unbranched structures contiguous with one another. In principle it would be possible to map the whole of the two genes using two or three factor crosses, but this would involve a great deal of labour. Benzer instead devised the technique of **deletion mapping** to enable new mutations to be located quickly. As well as point mutations which are characterised by their ability to undergo reverse mutation to give a wild-type allele once again, it is possible to get mutations which will not revert. These are also unable to recombine to give wild-type progeny with point mutations at a number of different sites. It appears that these mutations must have lost a section of the genetic material, that is to be deletion mutations. By testing for recombination with point mutations and with other deletion mutations, it is possible to determine the extent of the deletion involved in each mutation. In this way Benzer was able to build up a family of deletions, and to use these to define regions of the A and B gene. If figure 54 is referred to, it can be seen for example that section A_3 of the A gene can be defined as the section present in deletion 4, which is missing in deletions 1, 2, and 3.

Suppose a new *rIIA* mutant is isolated, and we wish to locate this. The new mutant is crossed to each of the standard deletion stocks in the normal way to test whether wild-type recombinants can be produced. Suppose that the mutant gives wild-type recombinants with deletion stocks 3, 4 and 5, but not with 1 and 2, it can be argued that the new mutation must affect a portion of the gene present in

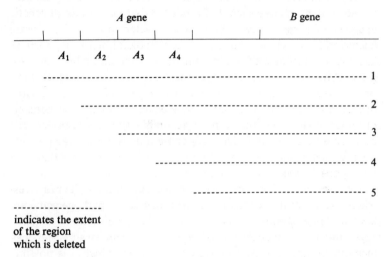

indicates the extent
of the region
which is deleted

Figure 54. (For explanation, see opposite.)

stocks 3, 4 and 5, but absent in stocks 1 and 2, i.e. the new mutation must be located in region A_2.

Once deletion mapping has given a rough location of the new mutation, conventional mapping can be used to characterise it more fully. In this way a high resolution map of the *rII* region has been built up, involving over 300 sites. An interesting fact to emerge out of this study is that all sites within a gene do not appear to be similarly mutable. Out of a sample of 1612 spontaneously-occurring *rII* mutants of different origin, over 500 were found to carry a mutation at an identical site. The remaining 1100 strains carried mutations involving 250 other different sites, many of which were only affected in one strain. This variation over a 500-fold range in mutability clearly indicates that all bases in the DNA are not equally likely to undergo mutational alteration. It is possible that the mutability of a base may be affected by the bases adjacent to it in the DNA chain. Another possibility is that the DNA has a more complex structure imposed on top of the basic double helical polynucleotide chain so that some bases are more exposed than others.

The genetic analysis of the *rII* region has been an invaluable foundation for many other investigators of the structure and function of the genetic material, and other examples of work which used the *rII* system will be described in later chapters.

Genetic analysis of bacterial viruses has not of course been confined to the virulent 'phages such as T4, nor for that matter has all genetic analysis of viruses been confined to bacteriophages. Temperate bacteriophage such as λ have also been studied intensively, particularly in connection with the investigation of the mechanism of lysogeny. Outside the bacteriophage, there are few viruses which have proved useful for genetic analysis. Although tobacco mosaic virus has been of considerable use in the elucidation of the mechanisms of mutation and gene function, no recombination has been detected between different strains. This is because it does not appear possible for the cells of the host plant to become infected by more than one virus particle which is a prerequisite for recombination to occur.

Some genetic analysis has been carried out on the influenza virus, where many strains differing in their degree of pathogenicity are known. These strains can be grown on the membranes of chicken eggs, and mixed inoculation of two non-virulent strains onto egg membranes can give rise to infected areas from which it is possible to isolate virulent recombination strains. However the whole process is technically rather difficult and progress is therefore slow. This work may be of importance however in the study of virulence and in establishing methods of control.

8. Cytoplasmic inheritance

So far in this book only the most usual patterns of inheritance have been described. However it is found, but by no means commonly, that some characters are not inherited in the way which has been described for the majority. In eukaryotes, the basis of many of these phenomena appears to be that the hereditary information responsible is not located in the nucleus but in the cytoplasm. Such inheritance is therefore called **cytoplasmic** or **extra-nuclear**.

In eukaryotes the gametes which fuse to form the zygote are often quite different in size, the female gamete, or **egg** being very much larger than the male gamete, or **sperm**. That the nuclei of the gametes resemble one another in size was taken earlier in this book as circumstantial evidence of their hereditary role (see page 31). However, if some genetical information is also carried in the cytoplasm, the unequal contribution of cytoplasm to the zygote by the two gametes should provide a way of detecting it. The technique used for this is the **reciprocal cross**. Suppose two strains of a diploid organism, A and B, are crossed together, first using A to supply the large female gametes, and B to provide the small male gametes. In this case the zygote, and hence the F_1 generation progeny will receive their nuclear genes equally from each parent, but their cytoplasm almost exclusively from strain A. If the cross is now done again, but using strain B to provide the female gametes, the F_1 progeny should be identical with respect to their nuclear genes, but their cytoplasm will this time be almost exclusively derived from strain B. Because of this, characters which are dependent upon hereditary information located in the cytoplasm, should show a different pattern of inheritance in the F_1 progeny obtained in the reciprocal crosses.

Before proceeding, it must be emphasised that not all differences in reciprocal crosses are necessarily the result of cytoplasmic inheritance. For example, if a Shetland pony and a Shire horse are crossed, when the Shetland is used as mother, the foal is very much the same size as pure Shetland foals. When the Shire is used as mother the foal is much bigger and resembles a pure Shire foal. However this is not due to cytoplasmic inheritance but instead to differences in the uterine environment in which the foals developed before birth. This can be shown to be so, firstly because in their subsequent development, the two types of hybrid foal become more and more alike, and secondly because there are no detectable genetic differences between the two hybrids upon further breeding. Differences between reciprocal crosses are also encountered where sex linked genes are involved. The characteristic pattern of inheritance which occurs in this case was described in Chapter 3 (see page 28).

As an illustration of what appears to be true cytoplasmic heredity the inheritance of pale green leaves in *Antirrhinum majus* can be cited. When the flowers of plants with pale green leaves are fertilised with pollen from normal dark green leaved plants, the F_1 progeny all have pale green leaves. When the reciprocal cross is made, the F_1 progeny have normal dark green leaves. Furthermore this breeding pattern is continued, the F_1 progeny behaving in crosses exactly like the parent strains (see figure 55). This pattern of inheritance is quite distinct from that normally encountered for single factor differences where no differences in reciprocal crosses occur (see figure 7, page 14), nor does it resemble sex-linked inheritance (see figure 15, page 29).

Similar non-nuclear inheritance can be detected for characters in a wide range of organisms. Many of the examples given in this chapter will involve micro-organisms, particularly protozoa, algae and fungi where analysis is often simplified, particularly as generation times are usually short.

The first detailed example involves the unicellular protozoan, *Paramecium aurelia*. The cells of *Paramecium* normally contain nuclei which are diploid; however these occasionally undergo meiosis to produce four daughter haploid nuclei, three of which abort. The nucleus which remains divides once by mitosis, and a cell containing two haploid nuclei is formed. Two such cells can pair and exchange one of their two haploid nuclei, and then separate, a process which is also called **conjugation**. Each cell now has one of its own nuclei, and one from the other cell. The two nuclei in each cell fuse, and the diploid state is restored. By appropriate techniques it is possible to ensure that during nuclear exchange, virtually no cytoplasm is exchanged. The inheritance of a nuclear gene is shown in figure 56. It is possible to obtain strains of *Paramecium*, called **Killer** strains which are able to kill normal sensitive strains by excreting a poison to which they are themselves immune. If the inheritance of the killer character is studied, it is found to behave as though it was determined by a non-nuclear gene. Thus the killer

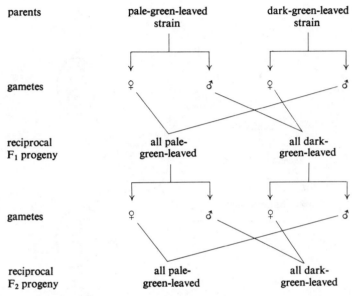

Figure 55. The inheritance of leaf pigmentation in *Antirrhinum majus*.

character is never transmitted with the nucleus, but instead all the progeny from the conjugating killer cell are killer, and all from the non killer cell are sensitive (see figure 56, where the cytoplasms have been shaded differently to indicate the heredity of a cytoplasmically determined character).

If the cytoplasm of a killer strain is examined, it is found to contain a large number of particles called **kappa**. Kappa particles are about 400 nm long, and contain DNA, resembling closely some types of viruses. Furthermore kappa particles can be shown to be infectious, as a sensitive strain can be converted to a killer strain by adding an extract of dead killer strain to its medium. It therefore seems likely that the killer character is due to the presence in the cytoplasm of a virus-like particle, which lives there in a symbiotic relationship with

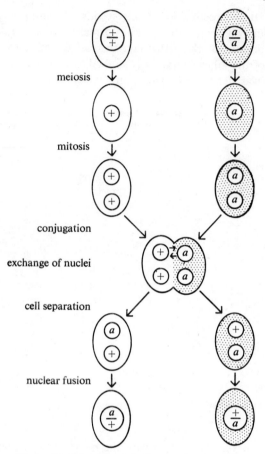

Figure 56. Inheritance in *Paramecium aurelia*.

the *Paramecium*. Other examples of cytoplasmic inheritance also seem to have a similar basis. In *Drosophila* the character CO_2 sensitivity can be explained by postulating the presence of an infective particle, **sigma**, in the cytoplasm of sensitive strains. Here however, no advantage is known to be conferred to sensitive strains and so the relationship seems to be parasitic, rather than symbiotic.

If these examples of cytoplasmic inheritance can be explained by the presence of a foreign organism in the cytoplasm, it can be asked whether all cases can be explained in this way. This question can be answered by investigating whether any natural cell components other than the nucleus, contain genetic information. Implicit in this question is the difficulty of answering which components of the cell are natural. However there is good evidence that two important cell organelles, mitochondria and chloroplasts do contain hereditary information, and this will be described before considering whether or not these organelles are natural cell components.

Cytologists have believed for some time that mitochondria and chloroplasts could not be formed *de novo* in cells and have described how they have observed these organelles dividing. It is difficult to obtain direct proof of this, but it has long been known that algae such as *Euglena gracilis* can lose their chloroplasts as a result of treatment with such agents as X-rays or ultra-violet light, and that this loss is irreversible. An extension of this work utilises an u.v. light microbeam and using this it is possible to show that it is the cytoplasm which is the sensitive target for inducing chloroplast loss. It has also been shown that ultraviolet light at a wavelength of 260 nm is maximally effective at inducing chloroplast loss. This last observation provided one of the first indications that DNA might be responsible for extra-nuclear inheritance too. Subsequently it has been shown that both chloroplasts and mitochondria contain DNA, and some strains showing inheritable cytoplasmic abnormalities can be shown to have altered organellar DNA (see below).

A great deal of work has been carried out on the inheritance of both chloroplasts and mitochondrial abnormalities. More perhaps has now been established about mitochondrial inheritance, and so this will be dealt with here in more detail. The yeast, *Saccharomyces cerevisiae* has proved to be ideal material for these investigations. Mitochondria contain most of the enzymes which catalyse the aerobic breakdown of sugars. Since yeast has a well-developed fermentation system it can survive even if its oxidative metabolism is impaired. It is possible to get mutant strains of yeast which are incapable of metabolising sugars oxidatively; such strains grow almost as well as the wild type on glucose, which they ferment, but on carbon sources such as acetate, which can only be metabolised oxidatively by the tricarboxylic acid cycle, they are unable to grow. These strains, called **petite**, can often be shown to be abnormal in their mitochondrial constituents. Some have certain mitochondrial enzymes and cytochromes absent, and others have the relative proportions of these components altered.

Before the inheritance of the petite character can be discussed, the life cycle of *Saccharomyces* must be considered. Although *Saccharomyces* and *Aspergillus nidulans* are fungi, in the same class, Ascomycetae; *Saccharomyces* has a life cycle rather different from *Aspergillus*. Normal yeast consists of single cells, each of which contains a single diploid nucleus. These cells multiply by mitosis. Under some conditions however these diploid cells undergo meiosis and each produces four haploid cells. These normally fuse in pairs almost at once, and so form diploid zygotes, which continue to divide by mitosis. Not any haploid cell can fuse with any other. In yeast there is a gene having two alleles A and α. All diploid yeast cells are heterozygous for these alleles $\left(\text{i.e. } \dfrac{A}{\alpha}\right)$. On meiosis A and α haploid cells will be produced in equal numbers, and A cells will only fuse with α cells, and *vice versa*. A slight complication that can be imposed upon the life cycle arises if the haploid cells are separated after formation. They will then divide by mitosis to give pure lines of haploid cells. Because these pure lines will have originated from a single cell, all cells in them will be either A or α, and so there will be no possibility of diploid formation. However if a haploid A culture is mixed with a haploid α culture, the cells from one culture will pair with the cells from the other, and zygotes will be formed. This life cycle is given in figure 57.

The behaviour of a cross involving a typical biochemical mutation, for example leading to a requirement for histidine, is shown in figure 58.

If the breeding behaviour of petite strains is studied, two quite different patterns emerge. Some petites, called **segregational** or **nuclear petites** behave in exactly the same way as the *his* mutation did in the above example. It is therefore probable that segregational petites carry a normal nuclear mutation, indicating that at least some of the information for mitochondrial structure is carried in the nucleus.

However some petites, called **cytoplasmic petites**, behave quite differently. In fact two extremes of behaviour are observed for cytoplasmic petites but all intermediate conditions are possible. One extreme is shown by **neutral petites**. When a haploid strain carrying a neutral petite is mixed with a normal haploid strain, the diploid strain is normal. Furthermore when this diploid strain undergoes meiosis no haploid petite strains are produced. With neutral petites therefore the petite mutation disappears on crossing (see figure 59).

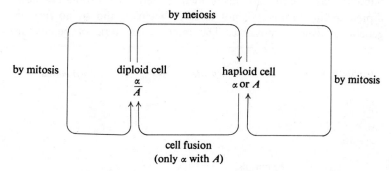

Figure 57. Life cycle of the yeast *Saccharomyces cerevisiae.*

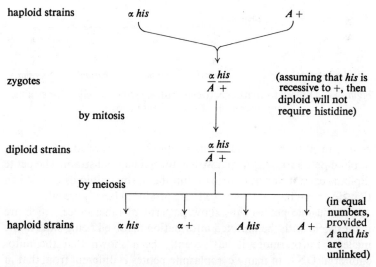

Figure 58. Diploid formation between a haploid yeast strain carrying a mutation in a nuclear gene (*his* = histidine requirement) and a wild-type haploid strain, and the result of subsequent meiosis.

The other extreme type of behaviour is shown by **suppressive petites**. If a haploid suppressive petite strain is mixed with a normal haploid strain, all the diploids formed are petite. Unfortunately these diploid petite strains will not undergo meiosis, and so no further genetic analysis is possible. The intermediate type of behaviour

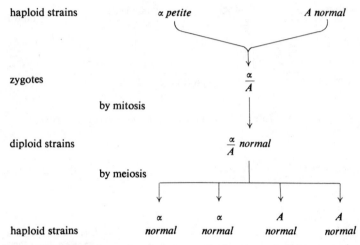

Figure 59. Diploid formation between neutral *petite* and wild-type strains, and the result of subsequent meiosis.

results in both petite and normal diploids being produced when a haploid petite strain is mixed with a normal haploid strain. The petite diploids cannot undergo meiosis, but the normal diploids can. When they do so, no haploid petite cells are produced (see figure 60).

Cytoplasmic petites thus show patterns of inheritance which are inconsistent with the mutated information involved being sited in the nucleus. Furthermore it has recently been shown that the mitochondrial DNA in many cytoplasmic petites is different from that in normal strains. In a few cytoplasmic petites, no mitochondrial DNA is even detectable. The abnormal segregation patterns coupled with the involvement of mitochondrial DNA, argue strongly in favour of some genetic information for mitochondrial structure being cytoplasmic, and probably residing as DNA within the mitochondria. The mechanics of mitochondrial inheritance appear complex however, as it appears that a diploid cell cannot tolerate a mixed population of mitochondria within it, and so when a haploid petite cell fuses with a normal haploid cell, one of the two mitochondrial types must be lost during subsequent mitosis. Furthermore the nature of the petite mutation itself affects which mitochondrial type is eliminated.

Further light has been shed on the situation as a result of the study

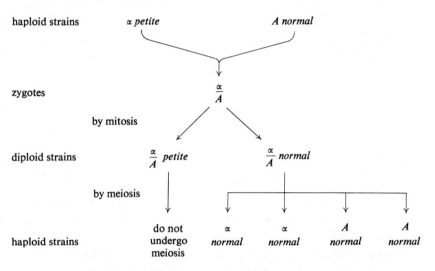

Figure 60. (For explanation, see opposite.)

of another class of yeast mutants which show abnormal behaviour in crosses. Certain antibiotics, among them chloramphenicol and erythromycin, inhibit cell metabolism by complexing with the ribosomes. These antibiotics are however more inhibitory to pro-karyotic cells than to eukaryotic cells, because they combine spe-cifically with the smaller (70S type) ribosomes found in prokaryotes. Although the majority of ribosomes in a eukaryotic cell are of the larger (80S) type, those within mitochondria and chloroplasts re-semble the typical prokaryotic ribosome. It is found that antibiotics of this type inhibit the aerobic growth of yeast, suggesting that they also interfere with mitochondrial ribosome function. It is possible to select yeast mutants which are resistant to these antibiotics, and when the behaviour of such mutants in crosses is studied, it is found to be similar to that of cytoplasmic petites. Again it appears that mitochondrial information is involved. Breeding experiments can now be carried out between various of these different mitochondrial mutants. When two different strains, one resistant for example to chloramphenicol and the other to erythromycin, are mixed, recombinant type diploid strains can be obtained (see figure 61). P. P. Slonimski and his co-workers have recently investigated the frequency of recombinant formation between various cytoplasmically

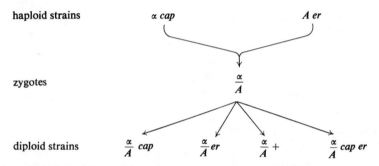

Figure 61. Diploid formation between haploid chloramphenicol-resistant (α *cap*) and erythromycin-resistant (*A er*) strains of yeast.

mutant yeast strains. The pattern which emerges is complex but has certain features which are reminiscent of recombination studies in *E. coli* before the introduction of the interrupted mating experiment. Thus Slonimski suggests that some mitochondria may be able to donate their DNA to others. Future developments in this field will be of considerable interest.

These cytoplasmic antibiotic-resistant strains can also be used to form diploids with cytoplasmic petites. When this is done it is generally found that petite diploids which arise are not antibiotic-resistant whereas the non-petite strains are. It therefore appears that whereas the genetic damage in the antibiotic-resistant strains cannot be too drastic, as different strains can undergo recombination, in petites the damage is more profound, and recombination with antibiotic-resistant strains cannot take place.

If experimental evidence leads us to the belief that some but by no means all of the genetic information for mitochondrial structure and function may be encoded in the mitochondrial DNA, it can be asked why it is necessary to have any of the information located there. Speculation on this point is rife. It has been suggested that because of the extremely lipophilic nature of some mitochondrial proteins, these must be synthesised *in situ*, and so the mitochondrion requires its own system of protein synthesis, distinct from the normal cytoplasmic one, so that these particular proteins can be synthesised in the right place.

Studies of chloroplast abnormalities are also in progress, and somewhat similar results are emerging. Thus recombination can occur in the alga, *Chlamydomonas* between different strains which appear to carry cytoplasmic mutations involving chloroplast function.

It seems therefore that both chloroplasts and mitochondria carry some of the genetic information necessary for their structure. This genetic information appears in both cases to be DNA and to be used in essentially the same way as the nuclear information. No new principles are therefore involved. The finding that these organelles have partial autonomy has given some weight to those who have proposed that mitochondria and chloroplasts may be extra-cellular in origin, representing extremely specialised prokaryotic symbionts. This theory is interesting but it is difficult to see how experimental proof to support it or dismiss it can be obtained.

9. Genes and proteins I: protein synthesis

Some of the evidence for the chemical nature of the hereditary material, and for the way this information is used by cells has been given in earlier chapters, but no detailed consideration of this process has been made so far. It will be recalled that in Chapter 1 (see page 4) the work of Beadle and Tatum was described. This lead to the proposition that the genetic information was used by cells to synthesise enzyme proteins. It has already been implied that other types of protein, including those involved in cell structure may also have their synthesis directed by genes. However the gene products are usually a little simpler than the protein itself. Many proteins consist of a number of polypeptide chains (chains of amino acids joined by peptide bands). These chains may all be identical, in which case the protein is said to be **homomeric**, or more than one type of chain may be involved, in which case the protein is said to be **heteromeric**. It is now thought that the basic product of a gene is a polypeptide chain, and so in the case of a heteromeric protein more than one gene is required to specify its structure. These slight complications to Beadle and Tatum's original theory are illustrated by considering the haemoglobin molecule. In humans, mutation in two different genes can affect the protein part of the haemoglobin molecule; which, it will be noted, is not strictly speaking an enzyme. Each haemoglobin molecule contains four polypeptide chains, two each of two sorts designated α and β. One gene determines the structure of the α polypeptide chain, and one the structure of the β polypeptide chain.

Another complication to a principle established earlier in this book can also be dealt with in this context. In Chapter 4 (see page 54), the complementation test was described, and it was stated that this test could be used to decide whether two mutations affect the same or different genes. If two mutations leading to the same phenotype could complement one another, it was concluded that these mutations must affect different genes. However the situation is sometimes more complex than this. It is found that it is possible to get mutations which whilst they are able to complement one another, are unable to complement further mutations. This sort of complementation pattern is best illustrated diagrammatically as in figure 62. In this example, it may be concluded that since strain 3 does not complement strains 1 or 2, they must all have mutations in the same gene. Similarly since

Mutant Strains	1	2	3	4
1	−	−	−	−
2	−	−	−	−
3	−	−	−	+
4	−	−	+	−

Figure 62. Typical pattern of complementation between four mutant strains. + indicates complementation, − no complementation between the two strains indicated at the top and side of the table.

strain 4 does not complement strains 1 or 2, these strains must also have mutations in the same gene. Since by crossing and reversion studies each strain can be shown to carry a single non-deletion type mutation, it must be concluded that strains 3 and 4 must be mutant in the same gene, even though they complement. The basis of this phenomenon which is called **intra-genic complementation** is now quite well understood. In each case that has been investigated, it has been found that the protein product is homomeric. The gene therefore directs the synthesis of polypeptide subunits, which complex to form the functional protein. Some mutations lead to the production of an altered polypeptide subunit, and in a few cases it is found that the homomer consisting of some subunits produced by one mutant allele, and some by another is at least partially active, even though the protein made entirely of one or other mutant subunit has no activity. As well as occurring within cells which contain both mutant genes (e.g. diploid cells or heterokaryons), this type of complementation also occurs *in vitro*. When the purified mutant gene products are mixed and treated appropriately, an enzymatically active hybrid protein can be produced. Furthermore by labelling the mutant products with radioactive isotopes, all the enzymatic activity can be shown to reside in hybrid molecules.

It will be seen that the product produced by intra-genic complementation is not the same as the wild-type product and it is usually quite easy to distinguish from it. Such characteristics as stability or catalytic activity are usually altered. The functional definition of a gene based on the complementation test can still be retained therefore, provided it is modified slightly. It can now be said that two mutations affect the same gene provided either they fail to complement or they complement to give a non wild-type product.

Having established that the genetic material is DNA, and for the most part located in the nucleus; and that the primary gene products are polypeptide chains, consideration can now be given as to how the cell uses its genetic information in protein synthesis. Cytological and biochemical studies of protein synthesis tell a great deal about this process, and some of the main conclusions from these types of studies will therefore be described here.

It can be shown that at least the great majority of proteins are not synthesised within the nucleus. One of the earliest studies to demonstrate this, was carried out in the 1930s by Hämmerling, who used algae of the genus *Acetabularia*. *Acetabularia* consists of a single large cell, containing for much of its life cycle a single nucleus, which is localised in the base of the cell. At the tip of the cell there is an umbrella-like structure referred to as a cap, and if this is removed, the cell will grow another one. The cells are sufficiently large to do intercell grafting experiments, and the basal region, containing the nucleus, can be grafted from one species onto an enucleated cell of another species. If the cap is subsequently removed the hybrid cell regenerates a cap characteristic of its nuclear component, thus providing still more evidence of the role of the nucleus in heredity. However, if the nucleus is removed from a young cell, this cell continues to grow for several months, producing a cap characteristic of normal cells of the species used. This was taken as evidence that the nucleus itself did not synthesise cap material directly, but that some intermediate carrier of information must exist. It seemed that this intermediate might be ribonucleic acid, RNA, the other form of nucleic acid found in cells. RNA differs from DNA chemically in two respects:

(a) there is a hydroxyl group on the 2 position of the sugar, i.e. the sugar is ribose not 2-deoxyribose,

(b) RNA contains the base uracil, but does not contain thymine.

RNA also differs physically from DNA in that it is single stranded. It is probable it can twist back on itself and form some hydrogen bonds between bases, but this is a very much more haphazard occurrence than in the double-stranded DNA, with its precisely paired bases.

RNA can be shown to be synthesised in the nucleus. The simplest way of demonstrating this is to observe the incorporation by the cell, of uracil or uridine (uracil + ribose) labelled with tritium (^{3}H). Since uracil is a component of RNA and not DNA, only the RNA will be labelled. If autoradiography is used it can be seen that the nucleus is first labelled with uracil, and that the RNA later migrates into the cytoplasm. Similar autoradiographical techniques using radioactively-labelled amino acids can be used to show that protein synthesis occurs in the cytoplasm. It can also be shown that the amount of protein being synthesised by a cell is correlated with the amount of RNA it contains, and in some cases where cells are able to take up the enzyme ribonuclease, which specifically destroys RNA, that this enzyme inhibits protein synthesis completely.

If a cell is fractionated, and the location of its RNA studied, it is found that most of the RNA is present in the ribosomes, small almost spherical organelles with a diameter of 20 nm, which consist of 60 % RNA and 40 % protein. Some RNA also occurs in fairly small molecules, with a molecular weight of about 25 000, and this fraction is often called **soluble RNA (sRNA)**. Even though both these types of RNA are, in eukaryotic organisms, synthesised in the nucleus, it does not seem likely for a number of reasons that either is the inter-mediate between DNA and protein synthesis. Firstly the base composition of these two RNA fractions is quite different from the base composition of the DNA of the species from which they are obtained. In fact it is found that the ribosomal RNAs from a wide variety of species have a rather similar base composition. The same is also true for soluble RNAs of different origins. However it is to be expected that the RNA which served as the intermediate carrier of information would be a copy of the DNA and would therefore have a base ratio mirroring that of the DNA. The second finding which counted against at least ribosomal RNA being the messenger was that there was no new ribosomal synthesis after 'phage infection of a bacterium, even though there was considerable synthesis of 'phage protein. Because of these discoveries a search was made for another RNA which would have the properties required of an intermediate. The first evidence for this third type of RNA was provided by further investigations of the infection of *Escherichia coli* by 'phage T2 and T4. First it was shown by A. D. Hershey, J. Dixon and M. Chase, that in the infected bacterial cell a small proportion of the RNA with a high turnover rate continued to be synthesised. In 1957, E. Volkin and L. Astrachan reported that they had labelled this fraction with ^{32}P, and by hydrolysis determined its base composition, which they found was close to the base composition of the infecting 'phage DNA; and quite distinct from that of either the bacterial DNA or of the majority of the RNA. A similar RNA fraction, with a high turnover rate, and a base composition close to the DNA, was later detected in uninfected *E. coli* cells, in yeast, and in the cells of a number of other species. This fraction which constitutes only about 2% of the total cell RNA was called **messenger RNA** (mRNA).

The next step in the elucidation of how messenger RNA functioned in protein synthesis involved the use of **pulse labelling** experiments. In such experiments a radioactively-labelled precursor of RNA or protein synthesis was added for a short period to a growing culture of cells, and was then effectively removed by adding excess non-labelled precursor, a process often referred to as **chasing**. The majority of experiments on protein synthesis have been conducted on *E. coli*, but the major findings have also been shown to hold for a variety of eukaryotic organisms. A growing *E. coli* culture is pulse labelled with ^{14}C-labelled uridine for 1–2 minutes, the cells are then broken open gently, and the cell components fractionated by density gradient centrifugation. Provided this process is carried out in the presence of 10 mM magnesium, it is found that the newly synthesised RNA (messenger RNA) sediments not only as though it had a molecular weight in the range 200 000 to 1 500 000, but also in association with ribosomes, and with particles of even larger size. Typical results of such an experiment are given in figure 63, where **svedburg units** (*S* units) are given as a measure of molecular size. These units are strictly speaking a measure of the speed of sedimentation of a molecule or particle in a gravitational field, and as such reflect not only its size but also its shape; however they are a widely used and convenient metric for this sort of work. If an electron micrograph is made of the heaviest fraction with which messenger RNA is associated, it is found to consist of clusters of ribosomes, joined by a thin strand. These are called **polysomes**. Mild treatment of polysomes with ribonuclease releases pulse-labelled material, and free single ribosomes,

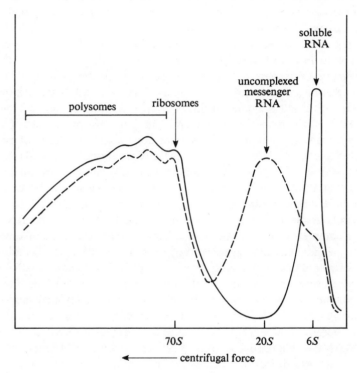

Figure 63. Fractionation by sucrose density gradient centrifugation in the presence of 10 mM magnesium of RNA from a growing culture of *Escherichia coli*. ——— total RNA (assayed by absorption of ultraviolet light at a wavelength of 260 nm.) - - - - - - pulse-labelled RNA.

suggesting that polysomes are groups of ribosomes held together by messenger RNA. If cells are now pulse-labelled with a radioactively-labelled amino acid the relationship of new protein synthesis to the RNA pattern can be determined. It is found that pulse-labelled protein material is also associated with the polysomes, and on ribonuclease treatment remains associated with the released ribosomes. Polypeptide synthesis must therefore take place on the ribosomes when they are associated with messenger RNA in poly-somes. The problem remains as to how the messenger RNA can achieve a precise control of the assembly of amino acids into poly-peptides.

The key to this problem, as will be seen below, is the soluble RNA. In the late 1950s several groups of workers had shown that soluble RNA could bind to amino acids. At about the same time, F. H. C. Crick had suggested that there might be some adaptor molecule involved in the process of protein synthesis which combined the property of recognising specific regions of nucleic acid with the recognition of particular amino acids. The soluble RNA appeared to be the ideal candidate for this adaptor role. If an amino acid and ATP are added to the soluble fraction from disrupted cells, it is found that the amino acid becomes complexed with part of the soluble RNA fraction. A detailed analysis of this process reveals that the soluble RNA fraction is heterogeneous, consisting of various types of RNA molecule, but all of about the same size. All contain from 70 to 80 bases, and have in common a cytosine–cytosine–adenine sequence at their 3′OH terminus. Each type of soluble RNA can complex with only one sort of amino acid, and each of these complexing reactions is catalysed by a specific enzyme. The complex is formed in two stages. First the enzyme, called an amino-acyl-sRNA synthetase or amino acid–sRNA ligase reacts with ATP and the amino acid to form an enzyme-bound amino-acyl adenylate complex (see figure 64). Then the enzyme-bound complex reacts with the appropriate soluble RNA, to produce an amino-acyl-sRNA complex and AMP (see figure 64). It is thought that the bases in sRNA molecules show some hydrogen bonding with one another. The complete base sequence of some sRNAs has now been determined and certain base sequences in different parts of the molecule are complementary to one another. If the molecule is arranged so that a maximum number of bases are paired, a clover leaf-like structure is obtained. The complete base sequence and possible hydrogen bonding for the alanine specific soluble RNA from yeast is given in figure 65. This complete base sequence, which was the first to be characterised, was determined by R. W. Holley and his co-workers, who published it in 1965. The sRNA molecules will not exist as flat molecules and the clover leaf structure must be thought of as being folded into a more globular form. It is almost certainly this complex tertiary structure of sRNAs which enables them to be recognised by the amino-acyl-

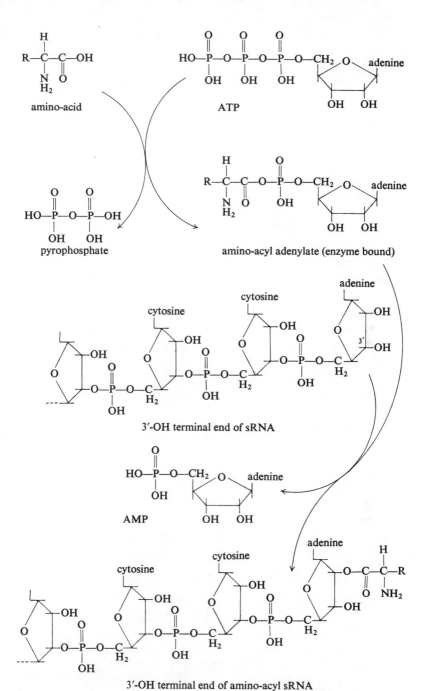

Figure 64. The reactions leading to the formation of amino-acyl sRNA.

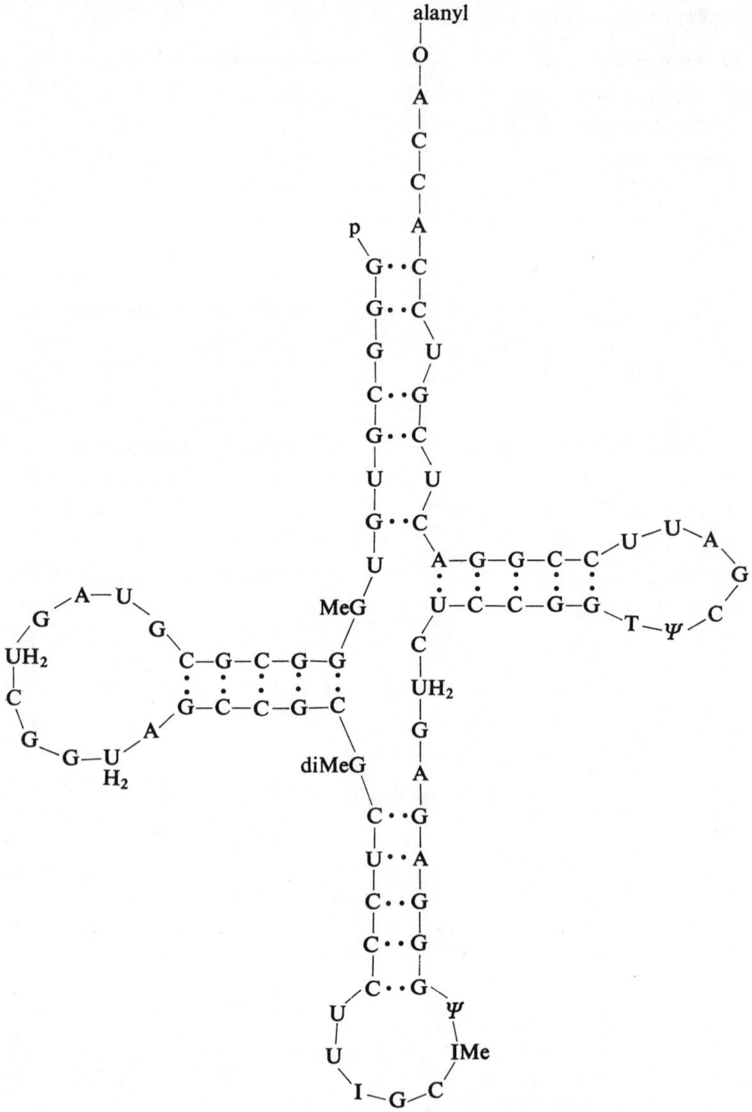

Figure 65. Base sequence and possible hydrogen bonding between bases for the alanyl sRNA from yeast.

Key: A   adenosine   IMe   1-methyl inosine
      C   cytodine   MeG   1-methyl guanosine
                                 T   ribothymidine
diMeG N^2 dimethyl guanosine   U   uridine
     G   guanosine UH_2 dihydrouridine
      I   inosine

Ψ pseudouridine

sRNA synthetase enzymes, and so ensure that they are always bonded to the correct amino acid.

The amino-acyl-sRNA complexes bind with the ribosome. Which particular amino-acyl-sRNA binds is determined by the messenger RNA which is also complexed with the ribosome. The details of this process are to be considered more fully below. Ribosomes consist of two subunits, unequal in size, the smaller subunit which for prokaryotic ribosomes sediments with a value of 30S, binds the mRNA, whereas the larger subunit (50S for prokaryotic ribosomes) binds the amino-acyl-sRNA complex. The 50S subunit however has two sites for binding charged sRNAs. One site, the **peptidyl site** will bind a peptidyl-sRNA complex and the other, the **amino-acyl site** will bind an amino-acyl-sRNA complex. The various sRNAs whose base sequences have been determined, have certain regions which show marked similarities. It is probable that these are the regions which are involved in the binding of the sRNA complexes to the 50S subunit of the ribosome. The sequence of events which is thought to occur during protein synthesis on the ribosome is as follows:

(1) The growing polypeptide chain is bound via an sRNA to the peptidyl site on the 50S subunit of the ribosome.

(2) The next amino acid to be added to the chain, becomes bound via its sRNA to the amino acyl site on the 50S ribosomal subunit. The specificity of this binding is determined by the mRNA bound to the 30S subunit.

(3) A peptide bond is synthesised between the polypeptide chain and the latest amino acid to arrive.

(4) The growing polypeptide chain is now attached to the ribosome via the sRNA of the last amino acid to arrive.

(5) The ribosome itself undergoes a change so that a new amino-acyl site is formed, possibly from the previous peptidyl site.

(6) as (1) etc.

In this way the ribosome moves along the mRNA and new amino acids are added to the peptide chain. A comprehensive diagram of this process which also takes into account further details to be described in the next chapter, is given later (see figure 74, page 153).

10. Genes and proteins II: the genetic code

Having considered the biochemistry of protein synthesis in broad outline, we can return to the problem of how DNA is able to control the precise structure of proteins. Clearly an obvious working hypothesis with which to start is suggested by the fact that both DNA and protein are linear polymers. This hypothesis proposes that it is the base sequence in DNA which determines the amino-acid sequence in protein. The earliest support for the hypothesis came from the detailed study of the effect of mutation on the protein produced by a gene. One of the first comprehensive studies of this kind was carried out by V. Ingram who investigated the structure of human haemoglobins. Man is not usually thought of as a suitable organism for the study of the molecular basis of genetics, but in this case the studies were extremely valuable. Ingram showed that the primary cause of a number of inherited diseases with symptoms of anaemia were changes in the structure of the haemoglobin molecule. He first showed that sickle-cell anaemia, which is caused by a recessive mutation inherited in the normal manner, resulted from the alteration of a single amino acid in the α-chain of haemoglobin. He went on to show that other anaemias with similar genetic bases, were all due to single amino-acid substitutions in the α or β-chain (see figure 73, page 152). Mutations have been found to have similar effects on proteins from a wide range of organisms. The enzyme protein, tryptophan synthetase, is one of the more intensively studied and the effect of mutation on this will be considered in detail below.

Such mutational evidence supported the hypothesis that there must be a special relationship between the sequence of bases in DNA and the sequence of amino acids in proteins. This special relationship became known as the **genetic code**. A great deal can be found out about the genetic code indirectly by the use of breeding experiments and these were carried out in parallel with the biochemical studies of protein synthesis which have already been described in the previous chapter. The synthesis of the findings from each of the fields of re-search to form a coherent model for the mechanism of gene action was a triumph for the joint biochemical and genetical approach to the problem.

The first problem about the nature of the genetic code that has to be decided is how the base sequence of DNA relates in a gross sense to the amino-acid sequence of protein. It had been assumed that the two were **colinear,** that is to say that if information occurred in the base sequence in a certain order, it was used in that order in deter-mining the sequence of amino acids in a polypeptide chain. This need not necessarily be so, and an alternative to colinearity is shown in figure 66.

The genetic code was not proved to be colinear until 1964, and then two groups of workers provided proof almost at the same time. One group led by S. Brenner used a slightly less straightforward approach, and the other proof will therefore be described here. C. Yanofsky and his research group have made an intensive study of the enzymes catalysing the synthesis of the amino acid tryptophan in *Escherichia coli*. One of the enzymes concerned, tryptophan synthetase, consists of two different polypeptide chains, *A* and *B*. Yanofsky studied part of the *A* chain, and determined the sequence of the amino acids in that region of the protein from wild-type strains. He then checked to see whether any of the large number of mutations which map in the gene determining the *A* chain's structure, affected the particular region whose sequence he had studied. He found 9 mutations each of which caused a single amino-acid change in this sequence. Details of the wild-type sequence, and the differences from it caused by these mutations is given in figure 67. He then carried out a careful linkage study of these 9 mutations, and established a linkage map. Details of this, including recombination fractions, are also given in figure 67.

Three important points emerge from this study.

(1) The order of the mutations established by genetic mapping corresponds to the order in the polypeptide chain of the amino acids affected by the mutations. This is direct proof of the colinearity of the genetic code.

(2) The genetic map distances (i.e. recombination fractions) reflect at least approximately, the physical distance between the affected amino acids in the polypeptide chain.

(3) It is possible to obtain recombination between mutations which affect the same amino acid. This is direct proof that the fundamental

COLINEAR CODE

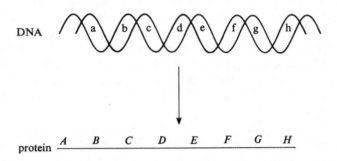

NON-COLINEAR CODE

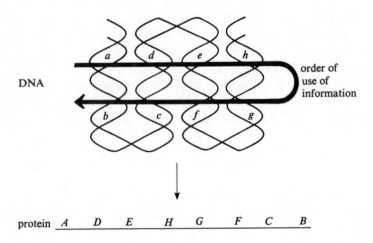

Figure 66. Alternative genetic codes.

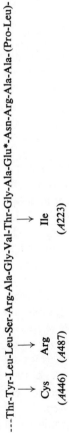
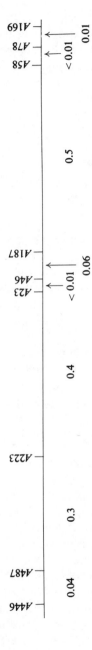

Amino acid sequence

---Thr-Tyr-Leu-Ser-Arg-Ala-Gly-Val-Thr-Gly-Ala-Glu*-Asn-Arg-Ala-Ala-(Pro-Leu)-

      ↑   ↑

     Cys   Arg   Ile

   (A446) (A487)   (A223)

Leu-Asn-His-Leu-(Ala-Val)-Lys-Leu-Lys-Gln-Tyr-Asn-Ala-Ala-Pro-Pro-Leu- Gln-Gly-Phe-Gly-Ser-Ile-Pro-(Asp⁺-Glu*)-Val-

                                         Arg Glu Val

                                     (A23) (A446) (A187)

Lys-Ala-Ile-Asp-Ala-Gly-Ala-Ala-Gly-Ala-Ile-Ser-Gly-Ser-Ala-Ile-Val-Lys-------

                ↙   ↓   ↘

               Asp   Cys   Leu

             (A58) (A78) (A169)

* possibly Gln
⁺ possibly Asn
sequences in brackets
have uncertain order

Linkage map

A446 — A487 —————— A223 ——————————————— A23 A46 — A187 ——————————————————— A58 A78 — A169

0.04 0.3   0.4   <0.01   0.5   <0.01

                                0.06   0.01

Figure 67. Yanofsky and his co-workers' data demonstrating the colinearity of the genetic code with the amino-acid sequence it determines. The upper part of the figure gives the amino-acid sequence of part of the *A* peptide of *E. coli* tryptophan synthetase, and the amino-acid alterations resulting from the various mutations studied. The lower part of the figure gives the linkage map for these mutations. Figures given are percent recombinations.

genetic unit, the base pair in the DNA must be smaller than the fundamental coding unit necessary to specify an amino acid. This fundamental coding unit is called a **codon**.

This last point is not surprising. As there are only 4 alternative base pairs which can occupy any one position in a DNA molecule, and there are 20 amino-acids which commonly occur in protein (see figure 68 for details of these), a base pair clearly contains insufficient information to act as a codon. In order to obtain the required information content, a codon must consist of several base pairs. Two adjacent base pairs are also not sufficient as only 4×4 alternatives exist. It can therefore be argued that a codon must consist of at least three adjacent base pairs.

This introduces another aspect of the genetic code which must be considered. Suppose each codon consists of three base pairs; there will then be 4^3 or 64 possible codons. But it has already been stated that there are only 20 amino acids to be coded for, and so either there is more than one codon for each amino acid, in which case the code would be said to be **degenerate**, or 44 codons would not code for amino acids, and would in other words be **nonsense**. Experimental evidence is given later in this chapter which indicates that the first alternative is nearer to the truth.

The next point to be decided about the genetic code is whether it is overlapping or not. Is each base pair in the DNA part of only one codon, or can some base pairs be involved in two adjacent codons? The various alternative possibilities are shown in figure 69.

There are two main objections to an overlapping genetic code. Firstly, if the code is overlapping it is to be expected that mutation in at least some bases will lead to alterations in two adjacent amino acids in the protein which is being coded. It is found that this is quite uncommon. The second objection is that since adjacent codons will share a base, this will impose some restrictions on which codons and hence which amino acids may lie adjacent to one another. Analyses of known amino-acid sequences in proteins have failed to reveal any such restrictions. It is possible to counter such objections by proposing that the coding ratio is greater than three, and then show that because of the additional redundancy created (if there were four base pairs in each codon there would be 4^4 or 256 codons to code for the 20 amino acids) mutation need not alter the amino acid coded for, and hence need not necessarily affect two amino acids. Such arguments were however overtaken by the evidence of other experiments which are to be described, and so it may be concluded that the genetic code is not an overlapping one.

It is now time to consider an extremely elegant series of genetic experiments, which established a number of important facts about the genetic code. These were carried out by F. H. C. Crick, L. Barnett, S. Brenner and R. J. Watts-Tobin who published their results in 1961. These workers used the *rII* system of bacteriophage T4. *rII* mutants have been described in Chapter 7 (page 106) and it will be recalled that whereas wild type 'phage give small fuzzy-edged plaques when grown on both strain *B* and strain K12 of *Escherichia coli*, an *rII* mutation leads to the production of large sharp-edged plaques on strain *B*, but prevents any growth on strain K12. The starting point of these studies was a 'phage strain which carried an *rII* mutation induced by the acridine proflavin, which mapped in the *B*1 section of the *rIIB* gene. The mutagenic action of acridines is to insert or delete single bases from the DNA (see Chapter 5, page 81).

Crick and his co-workers took this *rII* mutant strain and plated it on *E. coli* strain K12. The great majority of 'phage particles did not grow, but some plaques, formed by rare revertant strains, were obtained. The revertant strains were then plated on *E. coli* strain *B* where it was noted that many gave neither the typical *r* type plaque nor that typical of the wild type, but were instead intermediate, or **pseudowild**. Next, one of these pseudowild revertant strains was crossed to a wild-type 'phage strain, and a remarkable fact was established. It was found that among the progeny of this cross, were *rII* mutants identical to the original strain, and in addition a new type of *rII* strain, which also carried a mutation mapping in the *B*1 segment of the *rIIB* gene. It was concluded that the pseudowild phenotype resulted not from true reverse mutation at the original mutant site, but instead from the suppression of the original *rII* mutation by a further *rII* mutation in the same region of the *B* gene (see figure 70). This scheme was confirmed by crossing the original *rII* (*rII1*) mutant to the new one (*rII2*) and showing that wild type *and pseudowilds* were obtained among the progeny.

Since the suppressor of the original *rII* mutation is itself an *rII* mutation, a strain carrying it alone will be unable to grow on *E. coli* strain K12. It is therefore possible to select for revertants in this strain in the same way as was done in the original strain. When this was done exactly the same pattern emerged. The revertants were pseudowilds and it could be shown that a further mutation had

amino acid	abbreviation	structure	solubility in water (in g per 100 ml)
alanine	Ala	$H_3C-\underset{\underset{NH_2}{\mid}}{\overset{\overset{H}{\mid}}{C}}-COOH$	16.6
arginine	Arg	$H_2N-\underset{\underset{NH}{\parallel}}{C}-\underset{\underset{H}{\mid}}{N}-\underset{\underset{H}{\mid}}{\overset{\overset{H}{\mid}}{C}}-\underset{\underset{H}{\mid}}{\overset{\overset{H}{\mid}}{C}}-\underset{\underset{NH_2}{\mid}}{\overset{\overset{H}{\mid}}{C}}-COOH$	85.6 (HCl)
asparagine	Asn	$H_2N-\underset{\underset{O}{\parallel}}{C}-\underset{\underset{H}{\mid}}{\overset{\overset{H}{\mid}}{C}}-\underset{\underset{NH_2}{\mid}}{\overset{\overset{H}{\mid}}{C}}-COOH$	2.46
aspartic acid	Asp	$HO-\underset{\underset{O}{\parallel}}{C}-\underset{\underset{H}{\mid}}{\overset{\overset{H}{\mid}}{C}}-\underset{\underset{NH_2}{\mid}}{\overset{\overset{H}{\mid}}{C}}-COOH$	0.55
cysteine	Cys	$HS-\underset{\underset{H}{\mid}}{\overset{\overset{H}{\mid}}{C}}-\underset{\underset{NH_2}{\mid}}{\overset{\overset{H}{\mid}}{C}}-COOH$	very soluble
glutamic acid	Glu	$HO-\underset{\underset{O}{\parallel}}{C}-\underset{\underset{H}{\mid}}{\overset{\overset{H}{\mid}}{C}}-\underset{\underset{H}{\mid}}{\overset{\overset{H}{\mid}}{C}}-\underset{\underset{NH_2}{\mid}}{\overset{\overset{H}{\mid}}{C}}-COOH$	0.89
glutamine	Gln	$H_2N-\underset{\underset{O}{\parallel}}{C}-\underset{\underset{H}{\mid}}{\overset{\overset{H}{\mid}}{C}}-\underset{\underset{H}{\mid}}{\overset{\overset{H}{\mid}}{C}}-\underset{\underset{NH_2}{\mid}}{\overset{\overset{H}{\mid}}{C}}-COOH$	4.25
glycine	Gly	$H-\underset{\underset{NH_2}{\mid}}{\overset{\overset{H}{\mid}}{C}}-COOH$	25.3
histidine	His	$HC=\underset{\underset{HN}{}}{C}-\underset{\underset{H}{\mid}}{\overset{\overset{H}{\mid}}{C}}-\underset{\underset{NH_2}{\mid}}{\overset{\overset{H}{\mid}}{C}}-COOH$	4.16
isoleucine	Ile	$H_3C-\underset{\underset{H}{\mid}}{\overset{\overset{H}{\mid}}{C}}-\underset{\underset{CH_3}{\mid}}{C}-\underset{\underset{NH_2}{\mid}}{\overset{\overset{H}{\mid}}{C}}-COOH$	4.12

amino acid	abbreviation	structure	solubility in water (in g per 100 ml)
leucine	Leu	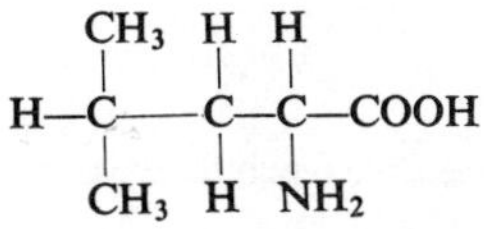	2.43
lysine	Lys	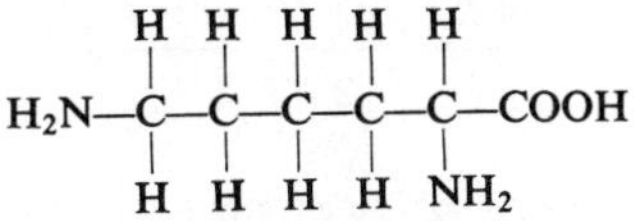	73.9 (HCl)
methionine	Met	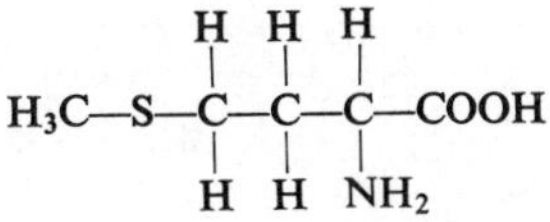	5.62
phenylalanine	Phe	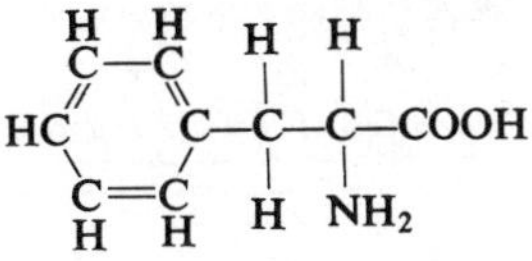	2.96
proline	Pro	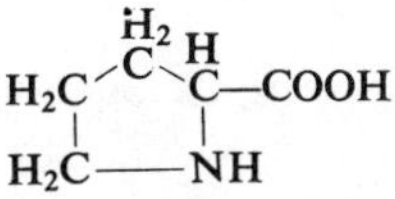	162
serine	Ser	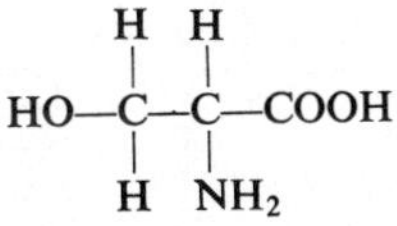	42.2
threonine	Thr	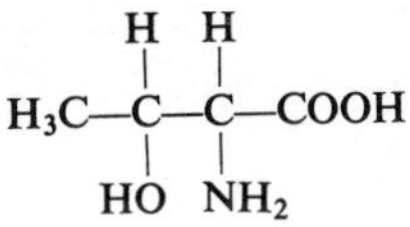	20.5 (DL-)
tryptophan	Try	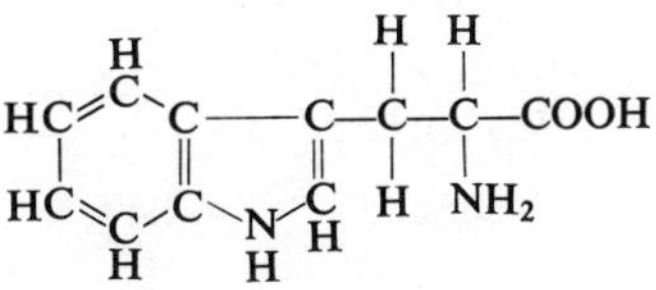	1.14
tyrosine	Tyr	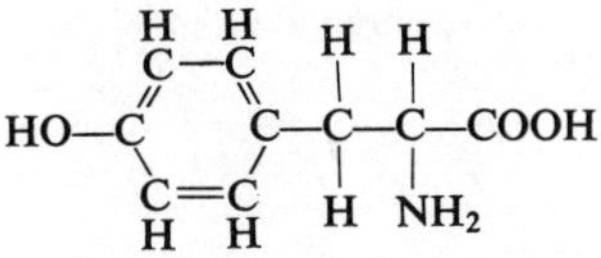	0.05
valine	Val	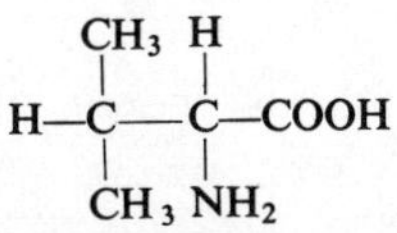	8.85

Figure 68. The structure and solubility of the twenty amino acids occurring commonly in proteins.

Non-overlapping code (each base pair is part of one codon only).

$$\underbrace{A\,B\,C}_{1}\,\underbrace{D\,E\,F}_{2}\,\underbrace{G\,H\,I}_{3}\,\underbrace{J\,K\,L}_{4}\,\underbrace{M\,N\,O}_{5}$$

Partially overlapping code (only some base pairs are part of two codons).

$$\underbrace{A\,B\,C}_{1}\,\underbrace{D}_{2}\,\underbrace{E\,F\,G}_{3}\,\underbrace{H\,I}_{4}\,\underbrace{J\,K}_{5}\,\underbrace{L}_{6}\,\underbrace{M\,N\,O}_{7}$$

Overlapping code (all base pairs are part of more than one codon).

$$A\,B\,C\,D\,E\,F\,G\,H\,I\,J\,K\,L\,M\,N\,O$$

Figure 69. Some types of genetic code.

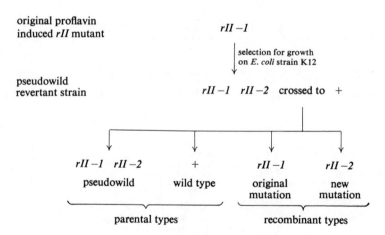

original proflavin
induced *rII* mutant   *rII −1*

                                 selection for growth
                                 on *E. coli* strain K12

pseudowild
revertant strain   *rII −1   rII −2*   crossed to   +

rII −1 rII −2   +   *rII −1*   *rII −2*

pseudowild   wild type   original   new
                                             mutation   mutation

parental types   recombinant types

Figure 70. (For explanation, see p. 137.)

occurred in the same region of the gene, which suppressed the second *rII* mutation. By taking the third mutation the process could be repeated, and then repeated once again, and so on. It was found that a particular *rII* mutation obtained in this way could suppress not only the mutation in the strain in which it was selected, but also a number of the other mutations obtained during the course of the series of experiments. The pattern that emerged was as follows. Suppose the original mutation is called a class 1 mutation, and those which were selected because they suppressed it, class 2. If mutations which suppress class 2 mutations are called class 3, and those which suppress class 3, are called class 4 and so on, then almost without exception a combination of two mutations one from an odd numbered class and one from an even numbered class gave a pseudowild phenotype. Two mutations both from even classes, or both from odd classes did not suppress one another, and none of these mutations was able to suppress mutations obtained by base analogue mutagens.

To explain these findings, Crick, Barnett, Brenner and Watts-Tobin made the following propositions (for the sake of clarity it will be assumed that a codon consists of three bases, a point established by this series of experiments, as will be seen below).

(1) The genetic information is read in sequence from a fixed point at one end of the gene. It is this fixed point that determines in which phase it is read, and without it there is no way of recognising how bases are grouped into codons. This process can be represented as follows.

$$A\ B\ C\ A\ B\ C\ A\ B\ C\ A\ B\ C\ A\ B\ C\ A\ B\ C\ A\ B\ C\ A\ B\ C$$

Fixed point   Codons

(2) The original mutation, being proflavin-induced was the result of the addition or deletion of a single base from the DNA. To simplify the argument it will be presumed to have been a deletion, but all that follows, holds equally were it an addition. As a result of this single base deletion, the genetic information would be read in the wrong phase once the mutated site has been passed.

B deleted

$$A\ B\ C\ A\ B\ C\ A\ C\ A\ B\ C\ A\ B\ C\ A\ B\ C\ A\ B\ C\ A\ B\ C$$

Fixed point   Read in wrong phase

The information read in the wrong phase will be incorrect, and so although only a single base is missing in the proflavin induced mutant, it will make a protein almost completely different from that of the wild type.

(3) The suppression of the original mutation was achieved by a further mutation close to it, which inserted an extra base in the DNA. As a result of this second mutation the reading of the genetic information was restored to the correct phase, once the second mutation was passed.

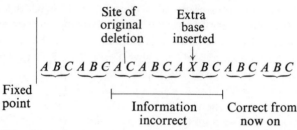

Only a small portion of the gene was read in the wrong phase, and the protein was therefore very similar to that in the wild type. As it was not identical however the suppressed strain might be pseudowild (rather than wild type).

(4) When the suppressor mutation was obtained on its own by crossing the suppressed strain to a wild type, it would, like the original mutation, cause the majority of the gene to be read out of phase.

(5) A similar argument can be extended to the whole series of suppressors, and it will be seen that it is to be expected that any addition (even class) mutation will suppress any deletion (odd class) mutation.

(6) It will also be seen that it is irrelevant to the general argument whether the original mutation was an addition or a deletion. If it was in fact an addition, all even class mutations would be deletions, and all odd class mutations, additions.

It is now possible to use this system to determine whether the code is read three bases at a time. If this is so then it should be possible to re-establish the correct phase of reading not only by the combination of an addition with a deletion, but also by the combination of three addition mutations or three deletion mutations in the same gene.

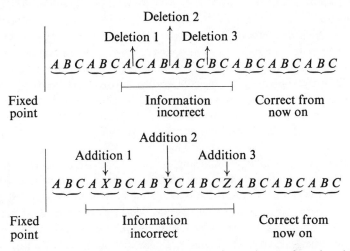

A number of such triple mutants were made, and were found to be pseudowild, which would be expected in view of the fact that the protein produced by such strains will not be identical to that of the wild type.

Crick and his co-workers showed therefore that the code had to be read from a reference point in order to make sense, and also that three bases coded for one amino acid. They also established another important point. Since a shift in the phase of reading will generate new codons randomly, and since nearly all addition mutations combine together with nearly all deletion mutations to give a pseudowild phenotype, nearly all codons must specify an amino acid; they therefore established by these experiments that the genetic code must be highly degenerate (see above, page 136).

It is possible to object to the conclusion which has been drawn from the experiments described above, that the code is based on triplets. Crick and his co-workers pointed out that since the exact way in which acridines cause mutation was not known, it might have been that their original mutant had an addition or deletion of more than one adjacent base. If this were true it would need a mutation which deleted or added the same number of bases in order to suppress it. The whole of the argument would still stand, except that it would be concluded that the coding ratio was six (for two bases added or deleted), nine (for three) etc. Strictly speaking therefore, it may only be concluded from these experiments that a codon consists of three or a multiple of three bases. It is possible to reject a coding ratio of nine on purely physical grounds. In certain small viruses, it is known that there is not enough DNA to code for the virus proteins produced if there were nine bases in each codon. However it has not yet been possible to determine the structure of all the proteins produced sufficiently accurately to reject a coding ratio of six on the same grounds. This can however be rejected by a different type of experimental procedure which involves a study of *in vitro* protein synthesis.

In 1961, M. W. Nirenberg and J. H. Matthei reported that they had devised an *in vitro* system which could synthesise protein. This system was complex, containing amino acids, amino-acyl-sRNA synthetases, sRNAs, ribosomes, ATP and various other necessary factors. When a synthetic RNA of defined structure was added to this, incorporation of amino acids into peptides occurred. By using a single amino acid labelled with a radioactive isotope, it was possible to see whether particular amino acids were incorporated into protein preferentially. The first result obtained was that polyuridylic acid (poly-U – the synthetic RNA containing uracil bases only), specifically stimulated the incorporation of the amino acid phenylalanine into protein material. This, in conjunction with the other experiments on the genetic code described in this chapter so far, indicated that the codon for phenylalanine must be three (or possibly six) adjacent uracils. This would be the messenger RNA codon, which would correspond to the DNA codon of three adjacent adenine–thymine base pairs.

Further elucidation of codons by this procedure was handicapped by the difficulty of synthesising RNAs containing bases in defined sequences for use as primers in the *in vitro* system. It was possible to get some idea of the likely codons by studying the pattern of amino-acid incorporation using an RNA containing various ratios of bases, but precise codon assignments could not be established in this way. H. G. Khorana and his team made some progress in this direction, and as a result of one experiment were able to decide whether the coding ratio was three or six. These workers were able to synthesise RNAs which contained two types of base only, and in which these bases alternated with one another. Predictions about the type of protein whose synthesis will be directed by such a co-polymer will differ depending on whether the coding ratio is three or six. If we consider the RNA containing the bases uracil and guanine only (poly-UG), if the coding ratio is three, it is to be expected that whichever base the RNA begins with, the same protein consisting of two alternating amino acids will be synthesised.

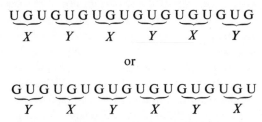

or

However if there are six bases in a codon, two different proteins each containing only one type of amino acid will be made.

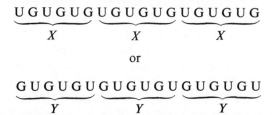

or

Khorana and his co-workers found that a protein consisting of alternating valine and cysteine amino-acid residues was synthesised, and so established that a codon must consist of an odd number of bases, which together with the evidence reviewed earlier indicates that the coding ratio must indeed be three.

A modification of the Nirenberg and Matthei original *in vitro* technique enabled most of the codon assignments to be established. Nirenberg and P. Leder found in 1964 that the addition of trinucleotides – RNAs only three bases long – to their *in vitro* system, whilst not stimulating any protein synthesis, caused amino acids to become bound to the ribosomes via their sRNAs. It is possible to make trinucleotides of precisely defined structure, and using these, they were able to determine which amino acids were bound. The procedure was simple. Twenty similar *in vitro* systems were set up, differing from one another only with respect to which of the twenty amino acids was labelled with radioactive isotope. To each was added the trinucleotide under study, and all were incubated. Following incubation each system was passed through a filter small enough to retain the ribosomes, but not small enough to hold back non-ribo-some-bound amino-acyl-sRNA complexes. Each filter was then examined, and it was usually found that in one of the twenty systems radioactive material had been retained on the filter. In this way the binding properties of each of the trinucleotides were examined, and in each case where codons had already been assigned as a result of previous experiments the trinucleotide corresponded to the established codon. It therefore seems that the trinucleotides function as though they were a small piece of messenger RNA, and that the system can be used to establish all the codons. It was found that no stimulation of binding could be established for a few codons, and the significance of some of these cases will be discussed below. A complete table of codon assignments, largely based on Nirenberg and Leder's work, but using other experiments as well, is given in figure 71.

If these codon assignments are studied the following points emerge:

(1) As was predicted from the experiments by Crick and his colleagues, the code is indeed degenerate, there being more than one codon for each amino acid except methionine and tryptophan.

(2) This degeneracy is not random. In 30 out of 32 cases it is not important which purine or which pyrimidine is present in the third position, and in 8 out of 16 cases it does not even matter which base is present. Crick has put forward a satisfactory hypothesis to account for this, but there is not space to describe it here (but see below, when sRNA – mRNA binding is discussed).

(3) The distribution of amino acids within the table does not appear to be random, thus the less soluble amino acids tend to occur to the upper left of the table, having uracil-rich codons, and the more soluble to the lower right, having guanine-rich codons. It has been suggested that this may have an evolutionary significance, for such an arrangement will tend to minimise the effect of mutation, many base changes resulting in either no change to the amino acid coded, or its substitution by one of similar structure.

(4) Not all codons code for amino acids. There are specific codons to indicate the point at which polypeptide synthesis should stop. The assignment of these codons was arrived at in a rather different manner which will be described below. There is also a codon which apparently has a double function, coding for the initiation of polypeptide synthesis as well as for an amino acid.

If the sRNAs act as adaptor molecules, recognising the codon of the amino acid they carry, then it is to be expected that they will have somewhere in their sequence, three adjacent bases which are complementary to that codon, and able therefore to specifically hydrogen bond to it. In all the sRNAs whose sequence has been analysed, this

		second base							
		U		C		A		G	
U	UUU	Phe	UCU	Ser	UAU	Tyr	UGU	Cys	U
	UUC	Phe	UCC	Ser	UAC	Tyr	UGC	Cys	C
	UUA	Leu	UCA	Ser	UAA	c.t.	UGA	c.t.	A
	UUG	Leu	UCG	Ser	UAG	c.t.	UGG	Try	G
C	CUU	Leu	CCU	Pro	CAU	His	CGU	Arg	U
	CUC	Leu	CCC	Pro	CAC	His	CGC	Arg	C
	CUA	Leu	CCA	Pro	CAA	Gln	CGA	Arg	A
	CUG	Leu	CCG	Pro	CAG	Gln	CGG	Arg	G
A	AUU	Ile	ACU	Thr	AAU	Asn	AGU	Ser	U
	AUC	Ile	ACC	Thr	AAC	Asn	AGC	Ser	C
	AUA	Ile	ACA	Thr	AAA	Lys	AGA	Arg	A
	AUG	Met*	ACG	Thr	AAG	Lys	AGG	Arg	G
G	GUU	Val	GCU	Ala	GAU	Asp	GGU	Gly	U
	GUC	Val	GCC	Ala	GAC	Asp	GGC	Gly	C
	GUA	Val	GCA	Ala	GAA	Glu	GGA	Gly	A
	GUG	Val	GCG	Ala	GAG	Glu	GGG	Gly	G

(first base, left margin; third base, right margin)

Key: U uracil   C cytosine   A adenine   G guanine
   For amino acid symbols see figure 68.
   c.t. these codons code for the termination of polypeptide chain synthesis.
   * this codon under some conditions codes for the initiation of poly-peptide chain synthesis.

Figure 71. The Genetic Code. Codons give the RNA trinucleotides coding for the amino acids indicated, and are written with the polarity 5′OH→3′OH. The table was compiled using as sources the work of Nirenberg and co-workers, Khorana and co-workers, Wittmann and co-workers, and Brenner and co-workers. It is therefore based largely on systems derived from *Escherichia coli*. The genetic code in other organisms is probably the same or very similar.

has been found to be true. This trio of bases is called the **anti-codon**. If the sRNAs are folded so that maximum base pairing occurs, then the anti-codon lies in an exposed loop where neither it nor the adjacent bases can hydrogen bond with other bases in the sRNA molecule. In figure 65 (page 130), the complete sequence for alanyl sRNA is given. From figure 71, it will be seen that the codons for alanine are GCU, GCC, GCA, or GCG. The anti-codons will therefore be CGA, CGG, CGU or CGC. The postulated anticodon is in fact CGI (I = nosine, the riboside of hypoxanthine – see figure 40, page 78 for structure). In DNA, it will be recalled (see page 79) that hypoxanthine paired, like guanine, with cytosine. It is thought that the codon–anti-codon pairing relationships between mRNA and sRNA cannot be so rigid as those in DNA. It is probable therefore that the CGI anticodon can bind not only with the GCC codon, but also with GCU and GCA, and this is the probable reason for much of the degeneracy found in the genetic code.

The establishment of which codons code for polypeptide chain termination was largely the work of S. Brenner and his co-workers. The existence in 'phage of a class of conditional lethal mutations, which were called *amber*, had been known for some time. 'Phage strains carrying amber mutations were characterised by being able to grow in some strains of bacteria (permissive hosts) whilst not being able to grow in others (non-permissive hosts). Amber mutations could be obtained in any 'phage gene, and Brenner's group, investigating amber mutations in the gene which specifies the head protein in 'phage T4, were able to show that in non-permissive hosts, only a protein fragment was produced. They argued that the amber mutation had led to a premature termination of polypeptide chain synthesis. They also studied the head protein produced by mutant strains in the permissive host, and showed that this was not identical to that of the wild type, having an amino-acid substitution at the site of the amber mutation. It had been shown that there was only a single gene difference between permissive and non-permissive strains of *E. coli*, and so Brenner argued that the suppression of amber mutations in the permissive host was due to its possession of an altered sRNA species, which now read the amber codon, and substituted its own amino acid in that position. Several different bacterial suppressor mutations are known, each substituting a different amino acid. It was found possible to obtain other mutations having similar properties to amber, but with different suppression patterns, and so there appeared to be more than one codon which coded for the termination of polypeptide chain synthesis. By a mutational analysis, using mutagens of known effects and noting to which amino acid codons chain termination codons could mutate, Brenner was able to deduce that the RNA code for the amber codon was UAG, and that there were two further chain termination codons, UAA and UGA.

Having established these codons, it can be asked whether they are recognised at the DNA to RNA (**transcription**) stage, or at the ribosome level when the RNA is **translated** into an amino-acid sequence. This question can be answered by an elegant experiment which is an excellent example of many sophisticated genetic experiments which can be performed using 'phage. First an amber mutation is obtained in for example the *B1* segment of the *rIIB* gene of 'phage T4. Continuing the convention used earlier in this chapter, we can represent this as follows.

Fixed
point

Chain
terminates

A strain carrying this amber mutation can then be crossed to a strain with a single base deletion, which is known to map near to the site of the amber mutation.

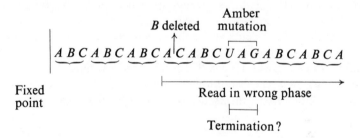

The UAG triplet if recognised at the transcription level will still be operative in bringing about chain termination. If it is recognised at the translation (mRNA→protein) stage however, it will no longer be recognised as the phase of reading has been changed. As the doubly-mutant strain is still mutant in phenotype it is not possible to determine at this stage, whether the amber triplet is still recognised. The next step is to cross the double mutant to a strain having a base addition mapping on the other side of the amber triplet from the deletion. This will restore the reading of the code to the correct phase, and unless the UAG triplet is recognised at transcription, the triple

mutant will be pseudowild. This is indeed found to be so. The triple
mutant can be represented as follows.

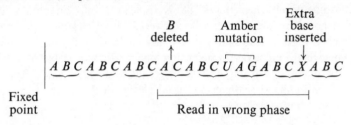

Chain termination triplets such as amber must therefore be operative
at the translation level. We can also ask whether there are signals in the
code which terminate RNA synthesis at the transcription level. It
will be seen in the next chapter that mRNAs often code for several
adjacent genes, indicating that the signal for the termination of RNA
synthesis must be more complex than the chain termination message
which marks the end of a gene. At the moment little is known about
these signals except that they must exist. Otherwise, for example, the
whole bacterial genome would be transcribed to give one enormous
RNA molecule. A corollary of this situation is that all genes would be
transcribed at the same rate, a situation which, as will be seen in the
next chapter, is known not to occur.

The last point of detail about the genetic code which will be dealt
with here, is the mechanism of polypeptide chain initiation. In *E. coli*
and its viruses, it was noticed that methionine was the most common
N-terminal amino acid in proteins, and that alanine and serine were
also common. Experiments using RNA from the 'phage *f2*, as a
primer in an *in vitro* protein synthesising system showed that whereas
in vivo the 'phage coat protein had a *N*-terminal sequence of ala-ser-
asn-phe-thr-, the *in vivo* product began *N*-formyl-methionyl-ala-
ser-asn-phe-thr-. It now appears that all proteins synthesised in
E. coli begin with *N*-formyl-methionine in the *N*-terminal position,
and that the formyl residue is then removed, and with it in some cases
the methionyl residue too. It has been shown that there are two
sRNAs which bind methionine in *E. coli*. Each is bound *in vitro* to the
ribosome in the presence of the AUG trinucleotide. *In vivo*, it is
found that the amino group of the methionine bound to one of the
two sRNA species is formylated enzymatically (see figure 72). It
appears that at the beginning of an mRNA, AUG codes for chain
initiation, binding the sRNA charged with *N*-formyl-methionine

whereas in the middle of an mRNA, AUG binds the 'normal' methionyl-sRNA. In fact it may be that the RNA sequence corresponding to met-ala-ser is required to act as the initiation signal. It is presumed that the formylation of the NH_2 group of the methionyl residue is necessary in order that it can be bound in the peptidyl site on the ribosome (see page 131), a prerequisite for the initiation of polypeptide synthesis. It is possible that AUG directs the insertion of N-formyl-methionine only when it is opposite the peptidyl site, thus allowing it to code for methionine when it occurs internally

$$
\begin{array}{ccc}
CH_3 & & CH_3 \\
| & & | \\
S & & S \\
| & & | \\
CH_2 & & CH_2 \\
| & \longrightarrow & | \\
CH_2 & & CH_2 \\
| & & | \\
H_2N-\underset{|}{\overset{}{C}}-COOH & & H-\underset{\parallel}{\overset{}{C}}-\underset{|}{\overset{}{N}}-\underset{|}{\overset{}{C}}-COOH \\
H & & O \quad H \quad H
\end{array}
$$

Figure 72. (For explanation, see opposite.)

within a polypeptide. In eukaryotes, the mechanism of polypeptide chain initiation is possibly different. It appears that although proteins are synthesised with methionine in the initial position, this is not formylated before incorporation.

It will have been seen that the majority of the properties of the genetic code which have been described in this chapter were elucidated as a result of experiments on either *Escherichia coli* or its viruses. There is however a great deal of evidence which indicates that the genetic code is substantially similar in all organisms. Perhaps the most direct comes from *in vitro* protein-synthesising systems. The same amino acids are incorporated into protein when a synthetic RNA primer is added to an *in vitro* system, whether this system is derived from a bacterial, plant or animal cell. Less direct, but *in vivo* evidence is provided by an analysis of the possible amino-acid changes caused by mutation. In tobacco mosaic virus, a plant virus which must utilise the protein-synthesising system of its host, the protein coats produced by a large number of mutant strains have been characterised, and in almost every case it is found that the amino-acid change could arise as a result of a change in a single base in the RNA. Similar results are found for known amino-acid changes in human

chain affected	amino acid position in chain	amino acid in wild type	amino acid in mutant	possible RNA codon change
α	16	Lys	Asp	AA_G^A GA_C^U
α	30	Glu	Gln	GA_G^A CA_G^A
α	57	Gly	Asp	GG_C^U GA_C^U
α	58	His	Tyr	CA_C^U UA_C^U
α	68	Asn	Lys	AA_G^A AA_C^U
β	6	Glu	Val	GA_G^A GU_G^A
β	6	Glu	Lys	GA_G^A AA_G^A
β	7	Glu	Gly	GA_G^A GG_G^A
β	26	Glu	Lys	GA_G^A AA_G^A
β	63	His	Tyr	CA_C^U UA_C^U
β	63	His	Arg	CA_C^U CG_C^U
β	67	Val	Glu	GU_G^A GA_G^A
β	125	Glu	Gln	GA_G^A CA_G^A

Figure 73. Amino-acid changes in human haemoglobin. In all except the first, it will be seen that the amino-acid change can be accounted for by a single base change in the nucleic acid.

haemoglobin. In figure 73 some of the most commonly known mutational alterations of haemoglobin are listed, together with the possible codon alteration which may have occurred.

Even if the genetic code is the same in all organisms, this does not mean that the codons for a particular amino acid are used with the same frequency. Indeed, if this were so, it would be difficult to explain how different species could show such wide variations in the base

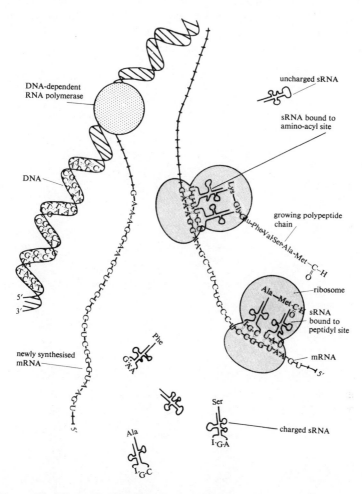

Figure 74. Schematic representation of the mechanism of gene action in *E. coli*. It has not been possible to draw the various components to the same scale.

composition of their DNA. It is likely that depending on the environment in which a species has evolved, some of the bases will be more readily available than others. The degeneracy of the code will allow the use of the most economic codons for an amino acid, without demanding a change in that amino acid. Hence another possible evolutionary significance of degeneracy is to allow for some flexibility in DNA structure, without affecting the encoded amino-acid sequence.

11. The control of gene activity I: intermediary metabolism

The way in which a cell uses its genetic information has now been described, but no mention has yet been made of whether genes are always active, or whether they are used by cells only when their products are required. It is only necessary to consider the diversity of cell types within a multicellular organism to conclude that selective use of genetic information must occur. Each cell, in for example a mammal, has come from a single zygote and this must therefore have contained the genetic information for the sum of all possible cellular processes. We are still a long way from understanding the mechanisms for such selective gene usage, but some simple mechanisms for the regulation of gene activity have now been characterised.

It is not perhaps surprising that these pioneer experiments have utilised *Escherichia coli*. It had been known from the end of the last century that many enzymes in micro-organisms could only be detected under certain conditions. Enzymes necessary for the fermentation of certain sugars for example were only detectable in cells which had been grown in the presence of that sugar. More recently another type of regulatory phenomenon had also been discovered. It was found that enzymes necessary for the synthesis of a certain biochemical, were often undetectable in cells which had been grown in the presence of an excess of the biochemical. In both cases it may have been alterations of enzyme activity only which had occurred, but even this would have effected cellular economies in the second case where possibly valuable metabolic intermediates would have been squandered unnecessarily had the synthesis of the biochemical been continued. However if enzyme synthesis itself was being controlled, then in both cases the economies would have been considerable, as not only is the process of protein synthesis energetically demanding, but also the component amino acids themselves must be obtained either in the diet, or by biosynthesis.

The first detailed investigation of these phenomena was initiated by F. Jacob and J. Monod, who investigated the metabolism of lactose by *E. coli*. The first step in the utilisation of lactose as a carbon source is its cleavage into galactose and glucose; catalysed by the enzyme **β-galactosidase**.

lactose   galactose   glucose

When the effect of the addition of lactose to a culture of *E. coli* is studied, it is found that before addition, and for a short period after addition very little of the enzyme β-galactosidase can be detected in the cells. Then after a lag of about 3 minutes in cultures grown at 37 °, enzyme activity begins to be detected, and almost at once the rate of appearance becomes constant. If lactose is removed from the culture, the accumulation of enzyme activity stops almost immediately (see figure 75). The first thing which had to be established was whether the appearance of β-galactosidase activity was due to *de novo* synthesis, or whether it came about by the activation of enzyme molecules already present. This can be decided by growing the bacteria in a medium containing radioactively-labelled amino acids, and then at the time of lactose addition, chasing the labelled amino acids with a large excess of unlabelled amino acids. When this was done, and the β-galactosidase produced was isolated, it was found to be substantially unlabelled, indicating that it must have been synthesised after the addition of the lactose.

It therefore appeared that the addition of lactose caused the genes necessary for the synthesis of β-galactosidase to become active very rapidly, and the removal of lactose was just as effective in inactivating these genes. Substances which cause the *de novo* synthesis of an enzyme are called **co-inducers**, or **effectors**, and are said to **induce** enzyme synthesis.

The next important finding that was made by Jacob and Monod, was that lactose was not the only substance which would induce β-galactosidase synthesis. A number of other galactosidases could act as co-inducers. However these were not necessarily substrates for β-galactosidase, and in fact some of the substrates which could be cleaved rapidly by the enzyme, were not very good co-inducers (see figure 76). Because of the differences in the spectrum of specificity of induction and substrate activity, it seemed likely that two different recognition sites must exist, one having to do with the induction

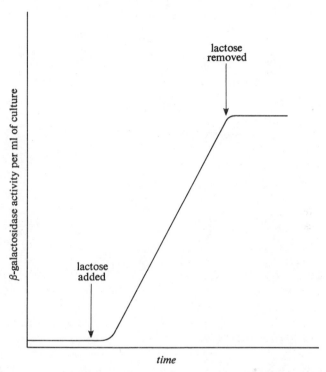

Figure 75. The effect of the addition and removal of lactose on β-galactosidase synthesis in a culture of *E. coli*. The time between the addition of lactose and the beginning of β-galactosidase synthesis is about 3 minutes in cultures grown at 37 °C.

mechanism, and the other being the catalytic site on the β-galactosidase molecule.

Jacob and Monod's next step in the investigation of the mechanism of the induction of β-galactosidase was to obtain mutations which affected this process. Mutations were already known which affected β-galactosidase directly. These latter mutations all mapped in one gene, z, which must specify the structure of the enzyme protein. This gene is therefore called a **structural gene**. Using z mutants which have lost β-galactosidase activity completely, it has more recently been established that, rather surprisingly, lactose itself cannot induce β-galactosidase synthesis. Thus when for example iso-propyl-β-D-thiogalactoside (IPTG) is added to a strain carrying a z mutation, it

	% induction	%$V_{max.}$
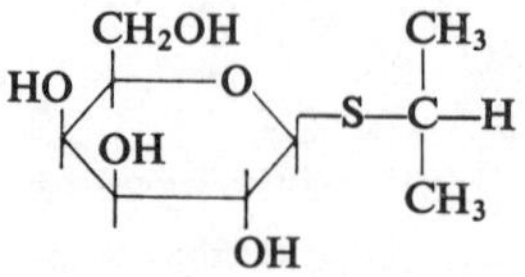 iso propyl β-D-thiogalactoside (IPTG)	100	0
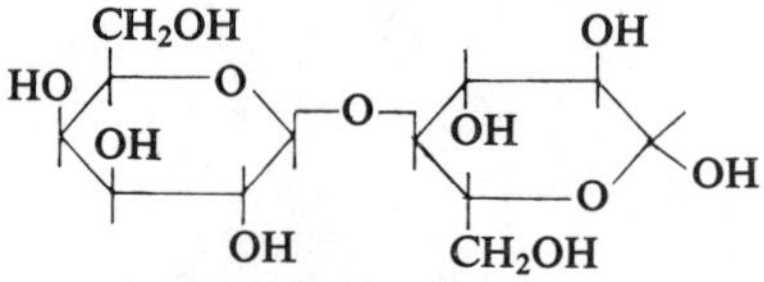 lactose	17	30
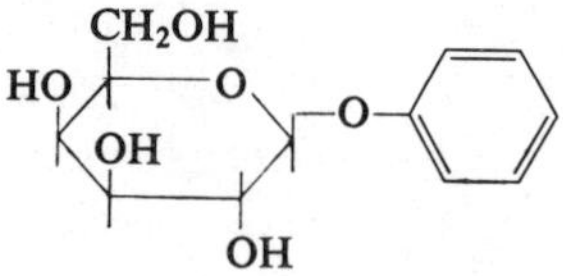 phenyl β-D-galactoside	15	100

Figure 76. Relative ability of three galactosides to induce and to act as substrates for the enzyme β-galactosidase. The maximum rate of reaction ($V_{max.}$) is used as the measure of the ability to act as a substrate.

can be shown that synthesis of the mutant gene product is induced in the normal way, but when lactose is added no synthesis occurs. It therefore seems that some product of β-galactosidase's action on lactose must act as co-inducer. As well as cleaving lactose to glucose and galactose, it is known that β-galactosidase catalyses the production of various di- and tri-saccharides, and it is now thought that one of these, produced by the low level of enzyme present in uninduced cells, induces enzyme synthesis. Because of this, it is better to use IPTG as a co-inducer especially as it cannot be cleaved by β-galactosidase. This means that there will be no change in the intracellular concentration of IPTG after induction. Such a co-inducer which is not a substrate of the induced enzyme is called a **gratuitous inducer**.

In contrast to the mutations in the z gene, it was also found possible to get mutations which did not affect β-galactosidase directly, but which altered the pattern of its synthesis, causing the enzyme to be produced even in the absence of a co-inducer. Strains carrying such mutations are said to produce enzyme **constitutively**. These constitutive mutations all mapped in the same region of the linkage map as the z gene, but they were found to be of two types. The first type which were called i^- were tested for dominance by obtaining F' bacteria (see page 95) which carried a mutant i^- gene on the F' agent, and the complete wild-type genome on the bacterial chromosome. The i^- allele was found to be recessive to the wild type (see lines 1, 2 and 3 in the table in figure 77).

Because the i^- allele is recessive to the wild-type allele, it may be concluded that the wild-type allele produces something, which the i^- allele cannot. Since the difference between wild type and i^- strains is that the former make no enzyme in the absence of co-inducer, we deduce that the something produced by the wild-type i gene must turn off enzyme synthesis under these conditions. Because of its mode of action the i gene product was given the name **repressor**. The i gene itself is called a **regulator gene**. It is possible to show that the repressor must be a substance which can travel through the cytoplasm, by constructing a partial diploid of the genotype

$$\frac{+^i\ z}{\text{F}'\ i^-\ +^z}.$$

In such a strain the only active z gene is in the same genome as the i^- gene which cannot make an active repressor. It can be shown that such a diploid has normally inducible synthesis of β-galactosidase, and so the repressor produced by the wild-type i gene must be able to travel through the cytoplasm and prevent the expression of the wild-type z gene when co-inducer is not present (see line 4 in figure 77). It is possible to get another class of mutations which map in the i gene. These are called i^s, and occur very rarely. They lead to the absence of β-galactosidase synthesis, and are dominant (see lines 5 and 6 in figure 77). The significance of the i^s mutation will be discussed below.

The second class of constitutive mutation found in this system maps in the o (**operator**) gene which is between the i and the z genes. These constitutive mutations are designated o^c. Dominance tests of o^c

	Genotype	β-galactosidase activity	
		Co-inducer absent	Co-inducer present
1	wild type	< 0.1	100
2	i^-	100	100
3	$\dfrac{+}{F'\,i^-}$	1	320*
4	$\dfrac{+^i\,z}{F'\,i^-\,+^z}$	1	240*
5	i^s	< 0.1	< 1
6	$\dfrac{+}{F'\,i^s}$	< 0.1	< 2
7	o^c	25	95
8	$\dfrac{+^o}{F'\,o^c}$	100	310*
9	$\dfrac{+^o\,z}{F'\,o^c\,+^z}$	100	220*
10	$\dfrac{+^o\,+^z}{F'\,o^c\,z}$	<0.1	100
11	$i^s\,o^c\,+^z$	45	105

*These high activities are explained by the presence of several F agents in each cell.

Figure 77. β-galactosidase activity present under inducing and non-inducing conditions in wild type and mutant strains of *Escherichia coli.*

mutations reveal a very significant fact. The o^c mutation is dominant but only with respect to the z gene which is adjacent to it. Thus a partial diploid of the genotype

$$\frac{+^o\ z}{\text{F}'\ o^c\ +^z}$$

synthesises β-galactosidase constitutively, whereas a diploid with a genotype

$$\frac{+^o\ +^z}{\text{F}'\ o^c\ z}$$

is still inducible (see lines 7, 8, 9 and 10 in figure 77). This is exactly the behaviour which would have been predicted if the o gene served as the site whereby the repressor recognised the z gene, and so controlled its expression. The o^c mutation would represent a change in the o gene which made it unrecognisable to the repressor, and so prevented the repressor's normal control of the z gene.

The model for gene control is completed if it is proposed that the co-inducer acts by complexing with the repressor, so that the repressor can no longer itself recognise the o gene. i^s mutations would lead to the production of a repressor with which the co-inducer did not complex, so that this repressor remained active even under inducing conditions. The finding that $i^s o^c$ double mutants are constitutive (see line 11 in figure 77) is consistent with this model, since although the repressor in such a strain would be active at all times, it would be unable to recognise the z gene because of the o^c mutation. The main features of this model are summarised in figure 78.

This model leaves some important questions to be answered such as the chemical nature of the repressor, and at what level, transcription or translation, gene activity is interfered with. Before going on to give the experimental evidence which enables their answer, one further complication of the lactose system must be introduced. There are two structural genes, a and y, adjacent to the z gene. The expression of these genes is controlled with the z gene, and so it appears that these three genes share a common operator (o) gene. The a gene specifies the enzyme thiogalactoside transacetylase, whose physiological role is unknown, and the y gene specifies the permease necessary for the transport of β-galactosides into the cell. This linked complex of genes is called an **operon**. There is good evidence that the operon is transcribed as a unit, and it is possible to isolate an mRNA which corresponds to the whole operon.

Returning to the question of the chemical nature of the repressor, this has now been established as protein. This had seemed likely as it had been known for example that it was possible to obtain amber-type i^- mutations. In 1966, W. Gilbert and B. Müller-Hill reported that they had been able to purify the lactose system repressor, using the technique of equilibrium dialysis. The repressor is a protein of

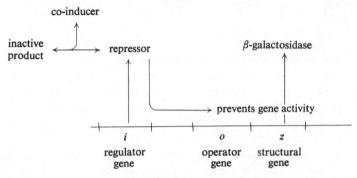

Figure 78. Model to explain the inducibility of β-galactosidase activity. β-galactosidase synthesis can occur for three reasons:
1. The repressor is inactivated by the co-inducer.
2. A mutation in the *i* gene prevents the production of an active repressor.
3. A mutation in the *o* gene prevents the recognition of the *z* gene by the repressor.

about 150 000 molecular weight, which binds tightly to co-inducers, such as IPTG.

The purification of the *i* gene product also enabled the other question, that of the level of gene activity at which regulation was operative, to be answered. Assays for the lactose operon mRNA which had been devised, had shown earlier that not only did the level of β-galactosidase increase on induction, but also the level of mRNA from the operon. This cannot be taken as rigorous proof that transcription is the primary target of the control system, as there might be some feedback system operative which prevented transcription, if the mRNA produced was not translated. However it has now been shown that purified *i* gene product has an affinity for the DNA of the operator region, provided co-inducer is not present, but no affinity for the corresponding mRNA, or for denatured single stranded DNA. Control in the lactose system therefore appears to be at the transcriptional level.

Another point of interest which has emerged from work of the *Escherichia coli* lactose system is the existence of a further region on the other side of the *o* gene from the *z*, *y* and *a* genes. This region is called the **promotor** (*p*). Mutations in the promotor region lead to a reduction in the rate of messenger synthesis from the whole operon. The promoter is thought to be the region for the attachment of the DNA-dependent RNA polymerase, the enzyme responsible for transcription. A great deal is becoming known about this enzyme. It is a heteromer consisting of four different types of polypeptide subunits. One type of subunit, **sigma**, does not affect the activity of the enzyme once it is bound to the DNA, but is required to initiate RNA synthesis in the promoter region. It has been suggested that if there were a number of different sigma subunits, each of which allowed the polymerase enzyme to bind with a different set of promoters, then a cell could, by controlling which sigma subunits were produced at any time, determine the broad spectrum of genes which were transcribable. Such a control system might be important in regulating the development of a multicellular organism.

Having described the mechanism which controls the genes of lactose metabolism in *E. coli*, we can now consider the extent to which other gene systems are controlled in a similar manner. No other *E. coli* system has been investigated in such detail, and in none is the mechanism of control so completely understood. However in general it appears that the synthesis of many enzymes may be controlled in a similar manner. There are however quite marked differences in detail, and one of the most common variations will now be described. At the beginning of this chapter, it was said that not only were there enzymes whose synthesis was induced by the addition of some biochemical to the medium, but also enzymes whose synthesis was curtailed by the addition of a biochemical. This phenomenon is called **enzyme repression**, and the enzymes of tryptophan synthesis will serve as an example of an enzyme system in *E. coli* which is controlled in this manner. *E. coli* makes the amino acid tryptophan in a number of steps, each requiring catalysis by a different enzyme. When cells are grown in a medium containing no tryptophan, these enzymes are synthesised, but when exogenous tryptophan is supplied, synthesis stops and the enzymes become diluted out by growth. This results in the very logical situation whereby bacteria stop making tryptophan for themselves when it is supplied exogenously. The genes which direct the synthesis of the enzymes catalysing tryptophan synthesis (structural genes) map in one small region of the *E. coli* linkage map, and since these enzymes have their synthesis regulated co-ordinately, it is probable that these genes are arranged in an operon. Mutation in any one of the structural genes can lead to an inability to make tryptophan, and hence a requirement for exogenous tryptophan. It is also possible as with the lactose system, to get mutations which alter the regulation of the synthesis of the tryptophan enzymes. In this case however mutation leads to the production of the enzymes even in the presence of tryptophan. Such mutations are said to cause **de-repressed enzyme synthesis,** and strains carrying them are often referred to as **de-repressed** (cf. constitutive) mutants. As with the lactose system it is found that the mutations which affect the control of the tryptophan structural genes, map in two separate genes. One class is just like the o^c mutations, mapping close to the structural genes, and only affecting the expression of the physically adjacent operon. The other class corresponds to the i^- mutations in the lactose system. They map in a different gene, and are recessive to the wild type. Thus partial diploids containing one allele of this de-repressed type, and one wild type allele show a wild-type, that is 'tryptophan repressed', pattern of control. Once again it can be argued that control is negative, that is that there is a regulator gene which makes a repressor whose function is to prevent the expression of the structural genes when not required. The difference from the lactose system however is that whereas there the repressor was normally active, and the effect of the co-inducer was to inactivate it, in this case it must be presumed that the regulator gene product is itself inactive, and the effect of tryptophan is to activate it. For this reason the product of regulator gene is called an **apo-repressor**, and tryptophan is called the **co-repressor** (cf. apo-enzyme + co-enzyme give active enzyme). This model for the regulation of the tryptophan synthesis is represented diagrammatically in figure 79.

There is a phenomenon related to repression, which can also control gene expression, but at a rather superficial level. In many biosynthetic enzymes, the product of the pathway not only represses the synthesis of the enzymes concerned, but also inhibits the activity of the first enzyme of the pathway. This inhibition is usually called **feedback** or **end-product inhibition.** This type of regulation can effect immediate economies in metabolites, whereas repression is seen as having a rather more long term effect. It is not surprising to find that the concentration of end product required to inhibit the first enzyme is usually slightly lower than that required to bring about repression, so feedback inhibition can be thought of as a less permanent control system than repression.

The next point to be considered is the extent to which control systems of the lactose type exist outside the bacteria. Not very much progress has been made in elucidating the mechanisms of genetic

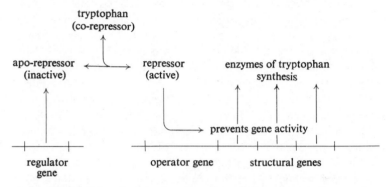

Figure 79. Model to explain the repression of the enzymes of tryptophan synthesis.

The enzymes will be synthesized for three reasons:
1. In the absence of tryptophan.
2. A mutation in the regulator gene prevents the production of an apo-repressor, and so repressor cannot be formed.
3. A mutation in the operator gene prevents the recognition of the structural genes by the repressor.

control in eukaryotes because they are in general very much less amenable to investigation than the bacteria. However a little progress has been made on the investigation of control systems in fungi, which lend themselves to analytical procedures similar to those employed with bacteria. Among the better understood fungal systems is nitrate reduction in *Aspergillus*, but as will be seen its analysis is not nearly as far advanced as that of the lactose system in *E. coli*. It is nevertheless an interesting system to consider, as it introduces several variations on the lactose control model, which are not however thought to be exclusive to fungi.

Nitrate must be reduced to ammonium before it can be incorporated into amino acids. In *Aspergillus* this reduction proceeds in two steps each catalysed by an enzyme. The first, the reduction of nitrate to nitrite involves the enzyme **nitrate reductase**, and the second, in which nitrite is reduced to ammonium, involves **nitrite reductase**. In the wild type, the synthesis of these enzymes is subject to control both by induction and repression. Grown on a nitrogen source such as urea, or an amino acid, only low levels of the two enzymes are produced. If nitrate is added to the medium however the synthesis of both enzymes is induced. If ammonium is now added, synthesis stops almost completely, even though nitrate may still be present. Typical enzyme levels under these conditions are given in the first line of the table in figure 80. As would be expected, there is a structural gene for each enzyme, and mutation in these may lead to a lack of the corresponding enzyme and hence an inability to utilise nitrate as a source of nitrogen. These structural genes although mapping fairly close to one another in the *Aspergillus* linkage map, do not appear to be arranged in an operon. Detailed analysis of the class of mutants lacking nitrate reductase (*nia*) reveals that most of these mutants synthesise the second enzyme in the pathway (nitrite reductase) constitutively (see line 2 in figure 80). Some *nia* mutants however have normally inducible nitrite reductase (see line 3 in figure 80). Mutations in the nitrite reductase structural gene (*nii*) do not affect the control of nitrate reductase synthesis (see line 4 in figure 80). All *nia* and *nii* mutants are subject to ammonium repression.

		Nitrate reductase			Nitrite reductase		
	Genotype	Urea	Nitrate	Nitrate + Ammonium	Urea	Nitrate	Nitrate + Ammonium
1	wild type	2	100	12	2	100	12
2	*nia* { majority	0	0	0	100	100	12
3	*nia* { minority	0	0	0	2	100	12
4	*nii*	2	100	12	0	0	0
5	*nir⁻*	2	2	0.5	2	2	0.5
6	*nir^c*	100	100	12	20	100	12
7	*mea^r*	2	100	100	2	100	100

Figure 80. Levels of the enzymes of nitrate reduction in wild type and mutant strains of *Aspergillus nidulans* grown on various nitrogen sources.

In addition to these classes of mutation, a third type exists which also leads to an inability to use nitrate as a nitrogen source. These map in a gene unlinked to *nia* or *nii*. Analysis of the enzyme levels found in strains carrying these mutations (*nir⁻*) shows that a low level of both nitrate and nitrite reductase is produced under non-inducing and inducing conditions. This level is the same as that produced by wild type under non-inducing conditions (see line 5 in figure 80).

It is also possible to get mutations affecting both aspects of regulation, induction and repression. Considering induction first, it is found that mutations occur very rarely which lead to the constitutive synthesis of both enzymes. These map in the same gene as the *nir⁻* mutations, and are designated *nirᶜ*. Strains having an *nirᶜ* mutation produce roughly the same amount of nitrate reductase in the absence of nitrate as they, or wild-type strains produce in its presence, but are still subject to ammonium repression. They also produce more nitrite reductase than the wild type when grown in the absence of nitrate, but under these conditions they do not produce as high a level of the enzyme as they, or wild-type strains do in the presence of nitrate, that is they still need nitrate for the maximum induction of nitrite reductase (see line 6 in figure 80).

Mutations leading to the de-repression of enzyme synthesis map in a fourth gene, unlinked to *nia*, *nii* and *nir*. Strains carrying such mutations are selected by their resistance to the ammonium analogue, methyl-ammonium, and so the mutant allele is designated *meaʳ*. Such strains although insensitive to ammonium repression, still require nitrate to induce the synthesis of nitrate and nitrite reductases (see line 7 in figure 80).

The first conclusion, that may be drawn from these findings, is that the processes of nitrate induction and ammonium repression appear to be independent. A similar situation is found for several bacterial systems. To gain a further insight into the likely regulatory mechanisms involved it is necessary as before to look at the dominance relationships of the various alleles which affect the control characteristics. Beginning with the alleles of the *nir* gene, which appears to be concerned with induction, *nir⁻* is recessive to both the wild-type and *nirᶜ* alleles, and the *nirᶜ* allele is dominant to the wild-type allele. This is therefore a complete reversal of the dominance relationships found for the various alleles of the regulator gene in the *E. coli* lactose system. There the constitutive (*i⁻*) allele was recessive to the wild type, and the non-inducible (*iˢ*) dominant. It can be argued that for the nitrate reductase system the opposite situation must hold, that is that the *nir* gene must produce a product which is needed for the expression of the structural genes. This regulator product would play a positive role, in contrast to those in the *E. coli* lactose and tryptophan systems. The simplest model, which it will be seen shortly needs further modification, would be that the wild-type *nir* gene makes a product, which is activated by nitrate, and is required for the expression of the structural genes. This product can therefore be called an **apo-inducer**, and nitrate would be a co-inducer. The *nir*⁻ mutation would lead to no functional apo-inducer being produced, and hence to no enzyme induction. The *nir*^{c} mutation would result in the production of an altered regulator product which would be active even in the absence of nitrate. The details of this simple model are given in figure 81. Since the *nia* and *nii* genes are not in an operon, two separate recognition sites would be involved. Mutation in these would lead to a loss in inducibility, and so would be indistinguishable from mutations in the structural genes.

This model does not explain why the majority of mutations in the *nia* gene lead not only to the absence of nitrate reductase, but also to the constitutive synthesis of nitrite reductase. Since the wild-type allele is dominant to the *nia* allele with respect to regulation, it can be argued that the wild-type product, that is the enzyme nitrate reductase must do something in addition to catalysing the reduction of nitrate to nitrite. Since most mutants unable to make nitrate reductase are constitutive, this extra function must be to prevent enzyme synthesis in the absence of nitrate. Mutation in the *nia* gene would usually abolish both functions, but could sometimes affect enzyme activity without abolishing the regulatory function. The converse type of mutation has not yet been found. The most plausible way of reconciling these new conclusions with the preliminary model, is to propose that the wild-type regulator gene produces an already active product (an **inducer**), which is inactivated by the enzyme nitrate reductase. When nitrate is present it will bind to the enzyme, which will no longer be able to inactivate the inducer. This will lead to the synthesis of enzyme until an equilibrium is achieved between nitrate concentration and enzyme level. This direct feedback by which the amount of the product of a gene controls the synthesis of that gene product has been called **autoregulation**. These new features of the control of the induction of the nitrate-reducing enzymes are included in the complete model, which is given in figure 82.

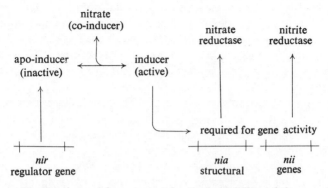

Figure 81. Simple model to account for the inducibility of the enzymes of nitrate reduction in *Aspergillus nidulans*.

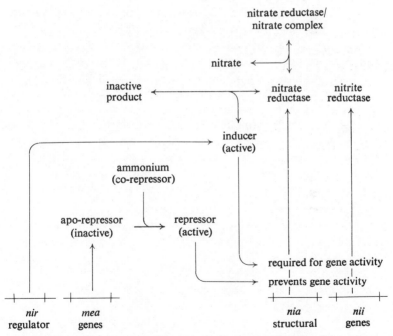

Figure 82. Model to account for the induction and repression of the enzymes of nitrate reduction in *Aspergillus nidulans*.

The mechanism of ammonium repression is less well characterised. It seems likely that the *mea* gene is a regulator gene similar to that described for the *E. coli* tryptophan synthetase system, and working in the same way. That is the direct product is an apo-repressor, and the ammonium ion acts as a co-repressor, activating it to form a repressor which prevents gene activity. These features are incorporated into the model given in figure 82.

There is no evidence at the present time for whether gene activity is controlled at the transcription or translation level in this system. The new features which emerge from the nitrate-reduction system are that control is positive, the structural genes are not grouped in an operon, there is both repression and induction, and finally a structural gene product is involved in regulation. None of these features is unique to this system, although no others combine all of them. Positive control, that is the requirement of a regulator product for gene activity, and a lack of operons do however seem to be more prevalent in fungi than in bacteria.

The systems of gene control which have been described in this chapter, have all involved enzymes concerned with intermediary metabolism. In all these cases control was reversible, that is the expression of a gene could be turned on, off and on again in response to alterations in the environment. In the next chapter some attempt will be made to assess the extent to which gene control in the course of the development of a multicellular organism is related to this metabolic control. It is tempting to think that the mechanisms involved will be similar, and many people have assumed that they are. However at the present time there is no evidence in favour of this, nor for that matter against it. A great deal of research effort is being put into this problem, and it is to be hoped that the next chapter will be the first in this book which will need to be rewritten.

12.   The control of gene activity II: development

The successes that have been encountered in the elucidation of the mechanisms controlling activity of genes, whose products are concerned with intermediary metabolism, were dependent in large measure on the firm biochemical foundation that existed in such systems. The structural genes were known, and their products, enzymes, could readily be detected and assayed quantitatively. It is largely because of a lack of this type of biochemical foundation, that it is proving so difficult to elucidate the mechanisms controlling the activity of genes which are concerned with the differential development of cells in a multicellular organism. The first part of this chapter will therefore be concerned with aspects of the study of gene action in development, and not necessarily with its control.

There is no difficulty in obtaining evidence that the differential development of cells is governed by genes. Mutations which affect the normal course of development have been known for many years. There are, for example, over eighty genes known in *Drosophila* in which mutation can occur so as to bring about alterations in the development of the wings. The difficulty arises when attempts are made to discover how any of these genes have their effect. One of the simpler facts which needs to be known is the time during the period of development, for which each gene is operative. There are various ways in which this can be established. Firstly, it should be possible to order the times at which various genes have their effects by studying the phenotype of strains mutant in two relevant genes. In general mutations in a gene which has its effect earlier in development should be epistatic to mutations in a gene which is operative later. It should in principle also be possible to study when a gene is operative in development, by the use of temperature-sensitive mutations. Suppose a mutation could be obtained which only affected the course of normal development at high temperatures. It would be possible to study the effect of keeping a mutant organism for only part of its development

171

at this high temperature, and the rest at a low temperature. In this way the stage of development at which the gene product was necessary could be determined, as this would be the only time at which high temperature would affect the mutant organism. Unfortunately this experimental approach in practice runs into difficulties. The body temperature of mammals is regulated by the organism itself, and it is not possible therefore to impose effective temperature changes on it. For organisms where temperature variation is possible such as *Drosophila*, these variations often have a profound effect on development even in the wild-type organism. Such techniques may also be feasible in plants, but no extensive use has yet been made of them there.

The effect of temperature on the development of a wild-type organism can however be used to study the problem of the time of action of genes. Such environmental effects as brief exposure to high temperature often mimic the effect of mutation. A wild-type organism which has been subjected to such a heat treatment may develop so that it resembles closely an organism with a particular mutation. These environmentally induced abnormal organisms are called **phenocopies**. By breeding from them, they can be shown to carry no permanent genetic defect. It can be argued therefore that if a mutation is mimicked by a heat shock given to a wild-type organism at a particular stage in its development, then both the mutation and the heat shock have had the same effect. It is likely that the heat shock has destroyed the efficacy of the product of the gene, which in the mutant strain had been inactivated by mutation. This then provides a way of discovering the times in which some genes are active in development, but this knowledge is not enough to decide how these genes bring about their effect.

There are many other approaches to the problem of studying gene action in development, some of which are more promising than others. Two will be described to serve as examples of the diversity of possible approaches. Some workers, recognising the difficulty in identifying the products of genes which are operative in development when so little is understood about the process, have decided to study the pattern of production of a well-known enzyme activity during the course of development. One such study by L. R. Nilsson has identified the different alkaline phosphatase activities produced by various stages of the *Drosophila* life cycle. Using the technique of poly-acrylamide gel electrophoresis which separates proteins not only according to their charge, but also to some extent according to their size and shape, a number of different enzyme proteins having alkaline phosphatase activity have been identified. These are not all present at all stages, some being present only for a short time during development, while others persist for longer (see figure 83). This clearly indicates that genes are controlled in development, and opens the way to obtaining mutants abnormal in their production of these enzymes, the study of which should add to our knowledge of the mechanism controlling their synthesis.

Another technique which is also used in *Drosophila* has the advantage of enabling gene activity to be studied within intact cells without knowing the gene product. In certain tissues of *Drosophila* the nuclei contain **giant chromosomes**, which consist of many replicas of normal chromosomes lying alongside one another. Because of this, it is possible to observe the detailed chromosome structure by the use of a microscope. After treatment with DNA-specific dyes, the chromosomes are seen to be banded, each band appearing to correspond to a gene or group of genes. At certain stages of development some of these bands are seen to swell up, and become more diffuse. These regions are called **Balbiani rings**, or more commonly **chromosome puffs**. By autoradiography they can be shown to be the site of intense RNA synthesis, and it is presumed therefore that a chromosome puff is a visible sign of an active gene. The pattern of chromosome puffing can be seen to vary not only from stage to stage in the same tissue, but also between tissues (see plate 2, page 175). Here again there should be an opportunity to study the mechanism regulating the activity of these genes, if mutations which affect these puffing patterns can be obtained. This system is also of interest because it provides an opportunity to study the effect of a hormone, some of which are known to have important effects on development. There was speculation, after Jacob and Monod published their model for the mechanism of the regulation of gene expression in the *E. coli* lactose system, that hormones might act as effectors by activating and inactivating repressors. How these hormones work is still not understood, but it is known that the insect hormone, **ecdysone**, a steroid, has an effect on puffing patterns in *Drosophila*, and so must bring about changes in gene activity. It is not at present known whether these effects are direct or indirect, but further work on this system may be able to answer this.

If then studies aimed at identifying the genes responsible for development and their products have not so far been very successful, is some other approach more revealing?

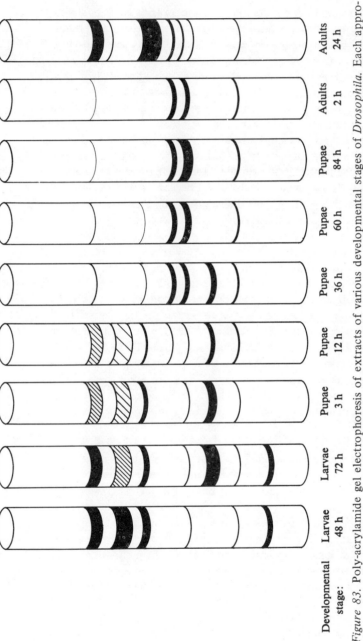

| Developmental stage: | Larvae 48 h | Larvae 72 h | Pupae 3 h | Pupae 12 h | Pupae 36 h | Pupae 60 h | Pupae 84 h | Adults 2 h | Adults 24 h |

Figure 83. Poly-acrylamide gel electrophoresis of extracts of various developmental stages of *Drosophila.* Each appropriate stage has been extracted in buffer, and the cell debris removed by centrifugation. The extract has then been placed on a small cylinder of buffer-saturated poly-acrylamide gel, contained in a tube, and an electrical potential has been established between the top and bottom of the gel. Charged proteins migrate down the gel at different rates depending on their charge, size and shape. When sufficient movement has taken place, the gel is removed from the tube, and the presence of proteins having alkaline phosphatase activity is detected by an assay in which this enzyme activity causes a pigment to be released which stains that region of the gel.

It is possible to speculate about the general nature of gene control in development and to answer some of the questions which arise. The studies already described go part way to answering a few of the

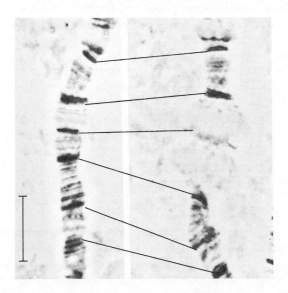

Plate 2. Giant chromosomes from the salivary gland of *Drosophila melanogaster*. The two preparations show the corresponding section of one of the giant chromosomes taken from two different developmental stages. The preparation on the right shows two large chromosome puffs which are not visible in the preparation on the left. (See page 173 for details.)

fundamental questions. If gene activity is to be controlled in development, this could be done in many ways. Four of the most important are as follows:

(1) Genes which are not required in a particular tissue may be eliminated in the cell divisions which gave rise to that tissue.

(2) Genes which are not required might be irreversibly blocked in some way, possibly by being complexed with a chemical, or by a mutation-like process.

(3) Gene activity may be controlled reversibly by processes similar to those described in the last chapter.

(4) Gene expression might be controlled by the activation or inhibition of the gene products.

A few general observations can be made about these possible methods of gene control. Firstly, method (1) and (2) are operationally similar, both are characterised by their irreversibility but it is possible that they might be distinguishable cytologically if gene loss was on a scale that could be observed. Methods (3) and (4) are both reversible systems, and so the transfer of a gene into an appropriate environment ought to result in its becoming active once again. The observation of chromosome puffing to some extent argues against method (4), but it must be remembered that several different systems of gene control could be operative even within the same organism.

We can now ask whether there is any evidence for the existence of any of these general mechanisms. Of mutation, little can be said, except that what is normally meant by mutation is a rare and rather haphazard phenomenon; and it is difficult, but not perhaps impossible, to see how mutation could bring about orderly development. Of gene loss, a little more can be said. It was stated above that if gene loss occurred on a large enough scale then it might be cytologically observable. Cytological observation of the tissues in the vast majority of organisms fails to reveal observable differences in the chromosome content of their constituent cells. In a few organisms there is however a form of orderly chromosome loss in some tissues. In both nematode worms, and in gall midges, there are some chromosomes (*E* chromosomes) which are present only in the **germ line**, i.e. those cells of the body which give rise to gametes. *E* chromosomes are eliminated from the cells of the **soma**, which constitutes the remainder of the body. In the gall midge, *Miastor*, there are 48 chromosomes in all, 36 of which are eliminated from the cells of the soma. It is not clear whether this chromosome elimination is the cause of the differences between soma and germ line, or a consequence of those differences. What can be said is that the phenomenon of chromosome elimination is certainly not widespread, and even where it does occur, it does not seem to occur in a way which is compatible with the idea that it is instrumental in differential cell development.

Another example of a rather different sort where differential chromosome content can be shown not be be responsible for differential development, is provided by the mosses. It will be recalled from Chapter 2 (see page 16) that the life cycle of many plants included two distinct stages, a diploid stage which gave rise to spores by meiosis, and a haploid stage which produced gametes. These two stages are called **sporophyte** and **gametophyte** respectively. In mosses, these two stages are morphologically quite distinct, the gametophyte being the leafy moss plant, and the sporophyte consisting of the stalk and round spore capsules which grow up from the gametophyte. It might be thought that the distinction between these stages was the result of their different chromosomal complement. However it can be shown that this is not so. If young sporophyte tissue is damaged, it can be made to regenerate into a typical gametophyte stage, which is however diploid not haploid. It is also possible, through mistakes in the fertilisation process to obtain haploid sporophytes. So the gametophyte does not owe its unique morphology to its haploid state, nor does the sporophyte to its diploid state, but instead other mechanisms, as yet unknown, must be operative.

It was stated above that proof that at least some of the gene control mechanisms involved in differential cell development were reversible, would be provided if an environment could be found in which inactive genes became active once again. Experiments aimed at investigating whether such environments could be found at first produced conflicting results, but the situation now appears to be a little clearer. One class of experiment, rather than studying the control of individual genes, attempted to find conditions in which a single differentiated cell could undergo further development to form new tissues, or perhaps a whole new organism.

In plants, it is possible to obtain, by damaging adult tissues, masses of rapidly dividing cells called **callus** which may be cultured indefinitely on a defined medium. The majority of the cells in callus are small and round; occasionally however cells are produced which are characteristic of differentiated plant tissues such as those which transport nutrients within the plant. Under certain conditions, it is also possible to get clumps of tissue and even shoots which can grow into whole plants. F. C. Steward has carried out a more careful study of this process. Starting with a callus culture obtained from a carrot root he obtained single cells, and showed that under appropriate conditions these could grow into complete new carrot plants, which were fertile. Such a cell which can give rise by division to a whole new organism is said to be **totipotent**. This experiment showed conclusively that at least in some of the cells in a carrot root, no genes can have been eliminated, or blocked irreversibly.

In animals, the amphibia have proved to be suitable material for investigating the totipotency of differentiated cells. R. Briggs and T. J. King conducted experiments on the American frog, *Rana pipiens*. In these, a nucleus is removed from a cell of some tissue, either adult or embryonic, by means of a micro-pipette. This nucleus is then transplanted into the cytoplasm of an egg cell, from which the nucleus has previously been removed by a similar technique. They argued that since the fertilised egg developed into the new organism, its cytoplasm should provide the appropriate environment for such development. They found that egg cells with transplanted nuclei did indeed undergo development, but they also found that donor nuclei obtained from progressively later stages in development, had a correspondingly less chance of developing after transplant. In fact, they only obtained a high rate of success when the nuclei were obtained from very early embryonic stages. As a result of these experiments, they argued that the processes involved in gene control in development must be irreversible, as nuclei from later developmental stages when transplanted into enucleated eggs seldom underwent normal development. However provided some nuclei were able to undergo extensive development, it is difficult to see how this conclusion is valid.

J. Gurdon extended Briggs and King's work. Using a different species of frog, the South African water toad, *Xenopus laevis*, Gurdon followed essentially the same procedure, except that instead of removing the egg nucleus mechanically, he destroyed it by ultraviolet irradiation. Gurdon has met with very much greater success, and he has for example obtained fertile adult frogs from transplanted nuclei taken from the cells lining the adult intestine. He suggested that Briggs and King's met with less success because *Rana* was not as tolerant as *Xenopus* of the experimental procedures employed. Critics of this type of work have pointed out that there is a considerable amount of DNA present in the cytoplasm of the amphibian egg, and that this might be responsible for development. However Gurdon has countered such criticism by transplanting nuclei from one strain of *Xenopus* into the cytoplasm of another strain, and showing that the individual which developed always resembled the strain donating the nucleus. The egg cytoplasm therefore appears to provide the environment alone, and these experiments give a clear indication that gene control in the development of Amphibia is, in all probability, reversible.

Another piece of work which sets out to study the reversibility of development in *Drosophila* indicates that there, quite complicated mechanisms may exist. *Drosophila* is a type of insect which undergoes development from an egg to a larva, then to a pupa and finally to an adult fly. In the larva there are distinct clumps of undifferentiated cells which during the normal course of development, are destined to give rise to certain specific structures in the adult fly. These clumps of cells are called **imaginal discs,** and there is, for example, one pair of discs which gives rise to much of the adult head, and another pair which give rise to the wings. E. Hadorn and his team of co-workers have dissected out these imaginal discs from a larva and transplanted them into the body cavity of an adult fly. There they remain alive, and proliferate. Provided they are transplanted into a new adult, before the old adult dies, imaginal disc cultures can be maintained indefinitely. If however the culture is transplanted into the body cavity of a larva, which then pupates and eventually becomes an adult fly, it is found that the cultured cells differentiate, and although complete organs are not formed, the cell types produced are recognisably characteristic of the various adult tissues. Hadorn has studied the types of adult tissue an imaginal disc culture can give rise to in this way, and has in addition studied the effect of the age of the culture on this process. For example, he might start with an imaginal disc which would be expected to form wing, and transplant this into the body cavity of an adult, where it would grow. The culture could be split, and part transplanted into another adult for continued maintenance, and the rest transplanted into a larva, which will complete its development. The range of cell types which arise from the latter transplant can then be determined. When this is done it is found that during the early stages of the maintenance of imaginal disc cultures these always give rise to cells of the same tissue type as would have been produced had the discs been left *in situ*. However after more prolonged culture (generally the passage of the culture through five successive adults is necessary), cells appear which are not characteristic of the tissue which would have been produced originally. It may be concluded therefore, that the cells in the larval imaginal discs, although apparently undifferentiated, are already programmed to give rise to certain specific tissues. Furthermore this programming is maintained even though these cells may undergo a number of divisions. However, eventually the programming breaks down, and other types of tissue may be produced.

A detailed study of this de-programming process shows that it cannot be random. For example, the first abnormal cell types produced by a culture of the imaginal discs which usually give rise to the adult genital apparatus, are characteristic of leg or antennal tissue. Next, cells may be produced which are characteristic of wing or labial (mouth part) tissue. Finally provided wing cells have been produced, typical eye cells may appear. Some cultures are not capable of giving rise to cells of certain types. For example, the discs which normally give rise to the legs, never, even after prolonged culture give rise to cells of the type found in the genital apparatus. Figure 84 summarises the various transitions which are found to occur.

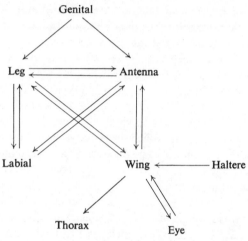

Figure 84. Structures which may arise from various imaginal disc cultures in *Drosophila melanogaster*.

Hadorn's work provides evidence once again that genes cannot be irreversibly switched off during the course of development, however he has also shown that getting genes switched on once again may be a difficult process, and may not in fact always be possible. Further investigations have shown that it is likely that cell division is necessary before de-programming can occur, and it is also known that mutations exist which mimic the effects of Hadorn's experimental procedures. There is for example a mutation *antennapedia*, which causes the adult fly to produce an additional pair of legs instead of antennae. Further study of such mutations, which could involve directly the mechanisms which control the differential development of the imaginal disc cells, may give us an insight into the nature of these mechanisms.

These various lines of investigations all indicate therefore that gene control in development is not irreversible. There are further experiments which give some indication of what general mechanisms might be involved in this control. It was suggested in the early 1950s that the basic proteins, **histones**, found associated with DNA in most eukaryotic chromosomes might have a regulatory role, and later, when Jacob and Monod published their findings, the idea was put forward that histones might be repressors. Evidence that this may be so, was provided by work on the pea, carried out by J. Bonner. Bonner studied the production of a globulin protein which has a storage function, and is made in appreciable quantities only in the pea cotyledons. He found that in apical bud tissue only about 0.1–0.2% of the protein was this globulin, whereas in cotyledon the globulin constituted from 5–9% of the total protein, depending on the age of the tissue. He next studied the ability of chromatin (DNA plus its associated proteins) isolated from apical bud tissue to direct protein synthesis in an *in vitro* system. He found that here again only about 0.1% of the protein produced was globulin. When the proteins, of which the majority are histones, were removed from the chromatin, and the DNA alone was used as the primer, the proportion of globulin produced was increased to about 0.5%. Bonner went on to show that DNA from cotyledon chromatin also produced globulin as about 0.5% of the total protein. The stripping of the proteins has therefore removed the organ-specific pattern of protein synthesis, and is consistent with the histones acting to prevent unwanted gene activity.

This experiment does not however provide conclusive proof of the regulatory role of histones. For example, it might be that histones had a purely structural function, and removing them so disrupted the structure of the chromosomes, that the normal control systems could no longer operate. More recently, experiments by J. Paul have indicated that this interpretation may be nearer to the truth. Paul has also studied the template activity of chromatin, which in this case is taken from various mammalian tissues. He does not attempt to obtain protein synthesis in his *in vitro* system but instead examines the RNA produced using a technique involving DNA–RNA hybridisation. There are certain criticisms which can be levelled at this method of RNA assay, but it is likely that the findings are for the most part valid. Paul has studied the pattern of RNA produced by chromatin *in vitro*, and compared it to that produced *in vivo* in the tissue from which the chromatin was obtained. He finds that there is a close correspondence between the two RNAs. If however, the protein is removed from the chromatin, the DNA produced specifies the synthesis of a much wider range of RNAs than before, confirming Bonner's findings with the pea. Paul has found however that chromosomal proteins can be added back to the DNA, and the chromatin reconstituted. The pattern of RNA synthesis directed by the reconstituted chromatin corresponds to the pattern obtained either using the original native chromatin, or *in vivo* in the cells from which the chromatin had been extracted. He next studied the effect of adding back only some of the chromosomal proteins. Chromosomal proteins can be separated into two major fractions; the larger being the basic histones, and the smaller consisting of acidic proteins. Paul found that if histones alone were added back to DNA, template activity was almost completely suppressed. If the acid proteins were alone added, there was little restriction of template activity compared with that of the protein-free DNA. However, when the acid proteins were added to the DNA followed by the histones, the original restricted, organ-specific pattern of RNA synthesis was once more obtained. Paul argued from these experiments that the role of histones is to mask DNA non-specifically, and so to prevent it from acting as a template for RNA synthesis. He proposed that the acid proteins functioned as the specific regulators, by preventing histones masking those genes which are required to be active. Undoubtedly more work must be done with this system before it can be certain that the histones do not play a specific role, but the results so far are certainly consistent with this being so.

The next general question that may be asked is whether the regulatory mechanisms responsible for the differential development of cells are positive, that is function by activating genes when their products are needed, or negative, that is function by keeping genes inactive except when their products are required. It is of course possible that both control modes may be used. H. Harris has attempted to answer this question at least in rather general terms. Harris has exploited the finding that some viruses cause cultured vertebrate cells to fuse and form multinucleate aggregates. Such viruses have this effect even after they have been inactivated with ultraviolet irradiation so as to render them incapable of causing infection. Harris has been able to induce cells of widely different origins to fuse and form heterokaryons, some of which are stable and undergo prolonged metabolic activity including cell division. In such heterokaryons the nuclei of different origins remain morphologically distinct and so continue to be recognisable. DNA and RNA synthesis by the nuclei can be detected using autoradiography after the cells have been cultured in tritium-labelled thymine or uracil. Harris has studied the effect of cell fusion on DNA and RNA synthesis by nuclei of different origins. By choosing cells with different characteristics, he is able to determine what type of controls must be effective. For example, HeLa cells, a malignant strain of human origin, are capable of both DNA and RNA synthesis. Chicken red blood cells on the other hand are capable of neither. Heterokaryons between the two cell types synthesise both DNA and RNA, and furthermore it can be seen that the nucleus of chicken origin has become capable of both syntheses. Further examples of this type of experiment are given in figure 85 and it will be seen that in each case the heterokaryon has the activities characteristic of its most active constituent. It can be argued from such results, that the control of RNA and DNA synthesis in these cells cannot involve a cytoplasmic inhibitor. On the other hand, it cannot be argued that control is necessarily positive, as the extra activities which are shown by some nuclei as a result of being placed in a heterokaryon, may be the result of some complementation-like process. For example, it might be that chicken red blood cells are unable to synthesise DNA, because the expression of some gene which directs the synthesis of an enzyme required for this process, is blocked. In the heterokaryon with a HeLa cell, the HeLa nucleus may be able to make an enzyme which will substitute for the one which cannot be made by the chicken red blood cell and so restore the latter's ability to synthesise DNA. It can only be said therefore that Harris's findings are not inconsistent with positive regulation mechanisms. They provide the basis for more experiments which should be able to investigate these mechanisms further. Consideration of this work should not perhaps be concluded, without commenting on the remarkable finding that nuclei of such different origins are able to co-exist in a fairly stable association, and that furthermore whatever the basis of observed effects, there must be a fair degree of similarity between the processes of gene expression in the various organisms.

The final topic which will be dealt with in this chapter is the nature of the triggers which cause the differential development of cells to proceed. If the two daughter cells formed by the division of a single cell develop differently, it is possible that this is brought about because the nuclei of those cells come to lie in different environments. The cytoplasm itself would be a very important component of that environment, and so the different developments of the cells might result from cytoplasmic heterogeneity in the original mother cell.

(a) *Pure lines*

	Synthesis of:	
	DNA	RNA
HeLa cells (human)	+	+
Rat lymphocyte cells	−	+
Rabbit macrophage cells	−	+
Chicken red blood cells	−	−

(b) *Heterokaryons*

Component A	Component B	DNA synthesis by component:		RNA synthesis by component:	
		A	B	A	B
HeLa	Rat lymphocyte	+	⊕	+	+
HeLa	Rabbit macrophage	+	⊕	+	+
Rat lymphocyte	Rabbit macrophage	−	−	+	+
HeLa	Chicken r.b.c.	+	⊕	+	⊕
Rat lymphocyte	Chicken r.b.c.	−	−	+	⊕

Figure 85. Patterns of synthesis of RNA and DNA by (a) pure lines of various cultured vertebrate cells and (b) heterokaryons between them. + indicates synthesis, − indicates no synthesis, and ⊕ indicates a synthesis of which that pure line is not capable.

There is a good deal of evidence that such heterogeneity exists, particularly in cells such as zygotes which must eventually generate a whole new organism. One of the simplest experiments which demonstrate the importance of the intracellular environment involves the development of pollen grains. Pollen grains are formed in tetrads as a result of meiosis, and these tetrads usually stick together for some time (see figure 86). Meiosis is followed by a mitosis in each grain which normally occurs so that the two nuclei are arranged radially with respect to the tetrad. The inner nucleus is destined to become the

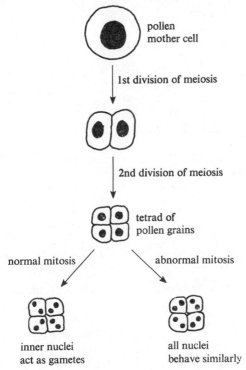

Figure 86. The early development of pollen grains.

gamete. However it is possible by various treatments to interfere with the plane of this mitotic cell division so that the two nuclei come to lie tangentially to the tetrad. When this occurs the two nuclei behave similarly but abnormally, neither functioning as a gamete (see figure 86).

Another experiment which provides evidence for cytoplasmic heterogeneity, involves the nematode worm, *Parascaris*. It will be recalled that nematodes are one of the groups of organism which show a selective loss of chromosomes in their somatic tissues (see page 176). This distinction between soma and germ line is established

at the first division of the zygote. The two cells produced are slightly
different in size, and the smaller loses chromosomes and gives rise to
the soma, whereas the larger shows no chromosome loss and gives
rise to the germ line. If the zygote is centrifuged however, it becomes
flattened and the plane of the first cell division is altered. Two similarly sized cells are produced, and neither loses chromosomes, with

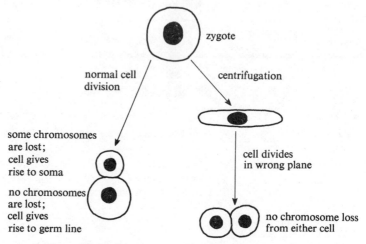

Figure 87. Early zygote development in Nematodes.

the result that subsequent development is abnormal (see figure 87).
It appears that centrifugation must have destroyed some cytoplasmic
heterogeneity which normally resulted in the differential development
of the daughter nuclei.

There are many other experiments which provide evidence for
cytoplasmic heterogeneity, but there are others which show that the
cytoplasm is not always necessarily important. For example, it is
possible to interfere with the plane of cell division in early stages of the
development of frog embryos without causing abnormal development.

Even if the cytoplasm is heterogeneous, it is still necessary to
explain how this heterogeneity originated. It is probable that the
activity of the nucleus is responsible, although the nucleus involved
would not be that of the affected cell, but instead those of the cells
around it, which in the case of the zygote would be those of the female

parent. In a developing embryo, it has been known for some time that the path of development of cells is influenced by substances produced by other cells. These substances are called **evocators**, and recently Tiedemann has characterised one, the **mesoderm inducing factor**, from an amphibian, the newt. This is a protein with a molecular weight of 25 000, but it is not known at present how it brings about its effect.

By now, it is to be hoped, some of the difficulties which are involved in the study of the mechanisms which control gene activity in development, will have been appreciated. The complexity of the process of development is itself one of the greatest difficulties, but other difficulties may be more conceptual in their nature. The mechanisms involved may be novel, and much of current research is aimed at the characterisation of developmental control mechanisms basically similar to those now known to regulate gene activity in microorganisms. It is perhaps appropriate therefore to conclude by pointing out that in at least one micro-organism, the protozoa *Paramecium*, there are mechanisms controlling development at the sub-cellular level which appear to have a hereditary basis not dependent on nucleic acid. T. Sonneborn has shown that the surface structures associated with the cilia covering the *Paramecium* cell have their orientation determined by that of pre-existing structures. This organisation can be disrupted by certain manipulations, and once this has occurred the abnormal pattern is inherited through many generations. Although the proteins involved in these structures are coded for by genes in the nucleus in the normal way, it seems that the assembly of these proteins into new units is determined by pre-existing protein arrays. It is not inconceivable that the cytoplasmic heterogeneity present in an egg cell could be due to this type of inheritance of cytoplasmic structure from egg to egg through the germ line.

Answers to problems

1. (a) Since the mother has blood group O, she must have the $\frac{I^O}{I^O}$ genotype, and her children must inherit an I^O allele from her. Their genotypes must therefore be

$$\frac{I^A}{I^O}, \frac{I^A}{I^O} \frac{I^B}{I^O} \text{ and } \frac{I^O}{I^O}.$$

Since a father can only contribute one of two alternative alleles, at least one of the children must be illegitimate. If he had blood group AB $\left(\text{genotype } \frac{I^A}{I^B}\right)$, then the child with blood group O must be illegitimate, if he had blood group A $\left(\text{genotype } \frac{I^A}{I^O}\right)$, then it would be the child with the B blood group, and so on.

(b) One of the children has blood group A, and another O; since the mother has blood group B, she must have the $\frac{I^B}{I^O}$ genotype. $\left(\frac{I^B}{I^B}\text{ homozygotes can have only B or AB children.}\right)$ The children must therefore have genotypes:

$$\frac{I^A}{I^O}, \frac{I^B}{I^O} \text{ or } \frac{I^B}{I^B}, \text{ and } \frac{I^O}{I^O},$$

and the father must therefore have either the $\frac{I^A}{I^B}$ or $\frac{I^A}{I^O}$ genotype. As the mother cannot taste PTC, she must have the $\dfrac{t}{t}$ genotype, and all her children must have inherited a t allele from her. Two children could taste PTC, and must therefore have the $\dfrac{T}{t}$ genotype while the third cannot, and must be $\dfrac{t}{t}$. The husband is therefore $\dfrac{T}{t}$.

Summary: mother: $\dfrac{I^B\ t}{I^O\ t}$; father: $\dfrac{I^A}{I^B}$ or $\dfrac{I^A\ T}{I^O\ t}$

children: $\dfrac{I^A\ T}{I^O\ t}$; $\dfrac{I^B}{I^B}$ or $\dfrac{I^B\ T}{I^O\ t}$ and $\dfrac{I^O\ t}{I^O\ t}$.

[handwritten annotation: — not possible due to 3rd child homozygous for O = O recessive blood type]

(c) The mother's genotype must be $\dfrac{I^O}{I^O}$, and each of her children must inherit an I^O allele. Their genotypes must therefore be

$$\dfrac{I^A}{I^O},\ \dfrac{I^B}{I^O} \text{ and } \dfrac{I^B}{I^O},$$

and their father's therefore $\dfrac{I^A}{I^B}$. Baby Y (blood group O, genotype $\dfrac{I^O}{I^O}$) cannot therefore be hers, as she and her husband can only produce children with blood of group A or B. The full genotypes of those involved must be as follows:

mother: $\dfrac{I^O\ m\ t}{I^O\ n\ t}$ $\qquad$ father: $\dfrac{I^A\ m\ T}{I^B\ n\ t}$ ✓

children: (a) $\dfrac{I^A\ m\ T}{I^O\ n\ t}$ $\qquad$ baby X: $\dfrac{I^A\ m\ t}{I^O\ m\ t}$

(b) $\dfrac{I^B\ n\ T}{I^O\ n\ t}$ $\qquad \left(\text{baby } Y: \dfrac{I^O\ m\ t}{I^O\ n\ t} \right).$

(c) $\dfrac{I^B\ m\ t}{I^O\ m\ t}$

(d) The genotype of the farmer is $\dfrac{m}{m}$, and his wife's $\dfrac{n}{n}$; and all their children must therefore have the $\dfrac{m}{n}$ genotype (blood group MN). The first son is therefore illegitimate. The second son could be legitimate. Full genotypes are as follows:

farmer: $\dfrac{I^O}{I^O}\dfrac{m}{m}$ farmer's wife: $\dfrac{I^A}{I^B}\dfrac{n}{n}$

second son: $\dfrac{I^B}{I^{O*}}\dfrac{m}{n}$ 'first son': $\dfrac{I^A}{I^{O*}}\dfrac{n}{n}$

2. Let the genotype of the true-breeding stock with colourless round seed be $\dfrac{col\ rou}{col\ rou}$. The unknown plant must contain information for red *and* colourless seed coat, and for round *and* square seed shape, i.e. must be heterozygous, because segregation occurs for these characters among the progeny. The unknown plant can be designated $\dfrac{+\ \ +}{col\ rou}$.

This will give the four classes of progeny when crossed to $\dfrac{col\ rou}{col\ rou}$:

$\dfrac{+\ \ rou}{col\ rou}$ red, round seed $\Big\}$

$\dfrac{col\ \ +}{col\ rou}$ colourless, square seed $\Big\}$ recombinant types

$\dfrac{col\ rou}{col\ rou}$ colourless, round seed $\Big\}$

$\dfrac{+\ \ +}{col\ rou}$ red, square seed $\Big\}$ parental types.

These four classes would occur with equal frequency if the genes were unlinked. The excess of parental types indicates linkage. The recombination fraction for the *col* and *rou* genes is $(60 \times 100)/5000 = 1.2\%$.

* Giving farmer benefit of the doubt.

3. The character wing-colour does not segregate independently of sex, and so sex-linkage is implicated. The fact that segregation occurs at all indicates the female fly is heterozygous (must contain information for yellow wings, but has grey wing, grey allele ($+$) dominant to yellow (y)). Cross is therefore:

$$\text{female } \frac{+}{y} \times \frac{+}{M} \text{ male.}$$

50% of the female progeny will be $\frac{+}{+}$, 50% $\frac{+}{y}$ and all will have grey wings. 50% of the male progeny will be $\frac{+}{M}$ (grey winged), 50% $\frac{y}{M}$ (yellow winged).

4. (a) Gene order. The gene order can be deduced without the calculation of recombination fractions. The least common classes of progeny will be those which are formed as a result of two crossovers. In cross (a) these are *paba y pro* and $+ + +$, and since the y and *pro* alleles have come from one parent, and the *paba* from the other, this gene order must be wrong (this order requires one crossover only to generate these classes). The order must instead be *pro-paba-y*, as this is the order which requires two crossovers to generate these least frequent types. In a similar way it can be deduced from cross (b) that the gene order is *paba-y-bi*.

(b) Recombination fractions, and genetic map.
In cross (a), the recombination fraction of *paba* and y is given by

$$\frac{(2 + 27 + 29 + 4)}{416} \times 100\% = 14.9\%.$$

The remaining recombination fractions can be calculated in a similar manner, and a map can be drawn as follows:

5. Since the F_1 progeny are tall, with normal leaves and green stems, tall (+) is dominant to dwarf (*dwf*), normal (+) to crinkly (*ckl*) and green (+) to brown (*bwn*). If it is assumed that each character is determined by a single gene (the least complicated hypothesis in view of the eight different phenotypes shown by the back-cross progeny), the genotypes of the two true-breeding strains would be

$$\frac{dwf\ ckl\ bwn}{dwf\ ckl\ bwn} \text{ and } \frac{+++}{+++}.$$

The F_1 will be $\dfrac{dwf\ ckl\ bwn}{+\ \ +\ \ +}$ and this when back-crossed to the

$\dfrac{dwf\ ckl\ bwn}{dwf\ ckl\ bwn}$ strain will generate eight classes of progeny as follows:

$\dfrac{+\ \ +\ bwn}{dwf\ ckl\ bwn}$ (201), $\dfrac{dwf\ +\ \ +}{dwf\ ckl\ bwn}$ (48), $\dfrac{dwf\ ckl\ \ +}{dwf\ ckl\ bwn}$ (196), $\dfrac{dwf\ +\ bwn}{dwf\ ckl\ bwn}$ (43),

$\dfrac{+\ +\ \ +}{dwf\ ckl\ bwn}$ (212), $\dfrac{+\ ckl\ bwn}{dwf\ ckl\ bwn}$ (50), $\dfrac{+\ ckl\ \ +}{dwf\ ckl\ bwn}$ (46), $\dfrac{dwf\ ckl\ bwn}{dwf\ ckl\ bwn}$ (206).

As these classes are not all equally frequent, linkage must be implicated. *bwn* does not appear to be linked to either *ckl* (486 recombinants out of 1002, recombination fraction $\simeq 50\%$) or to *dwf* (495 recombinants). *dwf* linked to *ckl*, recombination fraction = 18.7%.

6. (i) As progeny of the reciprocal F_1 cases are similar, no sex linkage is involved.

(ii) Parental stocks must contain information for blue and yellow eyes, even though they have white and green eyes.

(iii) As the back-cross progeny show four different phenotypes it is probable that two genes are involved.

(iv) Since these classes are equally frequent, there is no linkage between these two genes.

(v) The data are explicable if a stock homozygously mutant for both genes has white eyes, for one has yellow eyes, for the other has blue eyes, and for neither has green eyes.

Let the alleles be designated *yel* or +, *blu* or +.

Scheme:

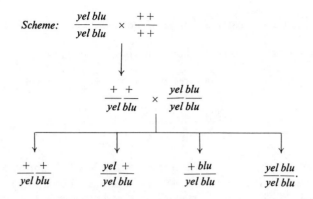

The biochemical basis might be as follows:

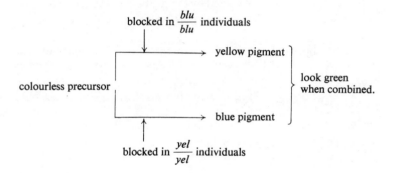

7. The F_2 ratio is near to 9 purple flowered : 3 red flowered ; 4 white flowered, which suggests a modified 9 : 3 : 3 : 1 ratio (i.e. two unlinked genes) where the double homozygous recessive is indistinguishable from one of the single homozygous recessive classes. This would occur if the gene whose mutant allele (w) leads to white pigment was epistatic to the gene whose mutant allele (r) led to red

pigment. The genotype of the white strain would be $\dfrac{w\ +}{w\ +}$, and of the

red strain $\dfrac{+\ r}{+\ r}$. Their F_1 would be $\dfrac{+\ r}{w\ +}$, and having a wild type allele for

each gene would be wild type (purple-flowered). Among the F_2

progeny, any $\dfrac{w}{w}$ types will have white flowers irrespective of their genotype for the r gene. The probable biochemical basis is:

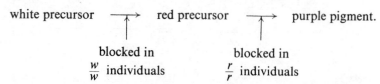

white precursor $\longrightarrow$ red precursor $\longrightarrow$ purple pigment.

blocked in $\dfrac{w}{w}$ individuals   blocked in $\dfrac{r}{r}$ individuals

8. As some of the F_2 progeny have white flowers, information for this character must be contained in the strains. These white strains account for about one sixteenth of the F_2 progeny, which again appears to be a modified $9:3:3:1$ ratio suggesting that two unlinked genes are involved and that only double homozygous recessive individuals have white flowers. If the mutant alleles of these two genes are designated w-A and w-B, and their wild type alleles $+$,

then the strains are $\dfrac{w\text{-}A\ +}{w\text{-}A\ +}$ and $\dfrac{+\ w\text{-}B}{+\ w\text{-}B}$, the F_1 $\dfrac{w\text{-}A\ +}{+\ \ w\text{-}B}$, and only the

$\dfrac{w\ A\ w\ B}{w\text{-}A\ w\text{-}B}$ F_2 progeny will have white flowers. Such a situation would

arise if the two genes were involved in alternative pathways for the synthesis of purple pigment as follows:

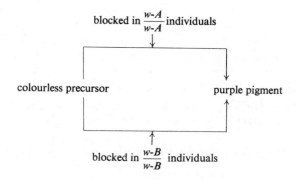

blocked in $\dfrac{w\text{-}A}{w\text{-}A}$ individuals

colourless precursor   purple pigment

blocked in $\dfrac{w\text{-}B}{w\text{-}B}$ individuals

9. (i) The segregation of hairy to hairless is the same irrespective of which flower-colour class is considered. This suggests that the genes determining these characters are unlinked.

(ii) The ratio of hairy to hairless is close to 3 : 1 among the back-cross progeny, which means that more than one gene must be involved in the determination of this character. Two genes would give a 1 : 1 : 1 : 1 ratio, which would be modified to a 3 : 1 ratio if only the double homozygous recessive progeny were hairless. If the mutant (hairless) alleles of the two genes are designated h-A and h-B and are recessive to their wild-type (hairy) alleles (+), then the two stocks are $\dfrac{+\ +}{+\ +}$ and $\dfrac{h\text{-}A\ h\text{-}B}{h\text{-}A\ h\text{-}B}$, and the F_1 $\dfrac{h\text{-}A\ h\text{-}B}{+\ \ +}$. Only $\dfrac{h\text{-}A\ h\text{-}B}{h\text{-}A\ h\text{-}B}$ individuals are hairless, and as in the last problem, it is probable that the two genes affect alternative functions.

(iii) The occurrence of four classes of flower colour among the back-cross progeny suggests that two genes determine this character also. The pattern of inheritance is basically similar to that determining eye-colour in problem 6. However since these classes are not all equally frequent, and since the parental types predominate, the two genes must be linked. The red- and blue-flowered progeny are recombinant types, and so the recombination fraction for the two genes involved is $(541 \times 100)/1533 = 35.3\%$.

10. (i) As there is segregation in the F_1, the two stocks must be heterozygous for a flower-colour gene.

(ii) As the stocks are true breeding, this heterozygosity must be concealed by epistasy.

(iii) As one quarter of the F_1 have red flowers, only two genes need be involved. Suppose that w-A and w-B are mutant alleles at two complementary genes, whose dominant wild-type alleles are designated +. One stock is homozygous for w-A, and segregates for w-B and +, whilst in the other stock the converse situation is found. The two plants will have genotypes

$$\frac{w\text{-}A\ \ +}{w\text{-}A\ w\text{-}B} \text{ and } \frac{+\ \ w\text{-}B}{w\text{-}A\ w\text{-}B}$$

and the F_1 will consist of four genotypes:

$$\frac{w\text{-}A\ \ +}{+\ \ w\text{-}B} \quad \frac{w\text{-}A\ \ +}{w\text{-}A\ w\text{-}B} \quad \frac{+\ \ w\text{-}B}{w\text{-}A\ w\text{-}B} \quad \frac{w\text{-}A\ w\text{-}B}{w\text{-}A\ w\text{-}B}.$$

Only the first class will have red flowers, and when selfed will give the 9 : 7 ratio typical of two unlinked complementary genes (see page 54). The possible F_2s are as in the table below.

	$\dfrac{w\text{-}A}{+}\ \dfrac{+}{w\text{-}B}$	$\dfrac{w\text{-}A}{w\text{-}A}\ \dfrac{+}{w\text{-}B}$	$\dfrac{w\text{-}A}{+}\ \dfrac{w\text{-}B}{w\text{-}B}$	$\dfrac{w\text{-}A}{w\text{-}A}\ \dfrac{w\text{-}B}{w\text{-}B}$
$\dfrac{w\text{-}A}{+}\ \dfrac{+}{w\text{-}B}$	$\tfrac{9}{16}$ red $\tfrac{7}{16}$ white	$\tfrac{3}{8}$ red $\tfrac{5}{8}$ white	$\tfrac{3}{8}$ red $\tfrac{5}{8}$ white	$\tfrac{1}{4}$ red $\tfrac{3}{4}$ white
$\dfrac{w\text{-}A}{w\text{-}A}\ \dfrac{+}{w\text{-}B}$		All white	$\tfrac{1}{4}$ red $\tfrac{3}{4}$ white	All white
$\dfrac{w\text{-}A}{+}\ \dfrac{w\text{-}B}{w\text{-}B}$			All white	All white
$\dfrac{w\text{-}A}{w\text{-}A}\ \dfrac{w\text{-}B}{w\text{-}B}$				All white

11. As with the last problem, the segregation in the F_1s suggests heterozygosity, which must again be concealed by epistasy. There are three types of mating.

(i) First pair give rise to no miniature-winged progeny; and as the F_2 show a segregation of 3 vestigial (vg) : 1 wild-type (+) a single gene appears to be involved. There is no evidence for sex linkage:

$$\frac{vg}{vg} \times \frac{+}{+}$$
$$\downarrow$$

F_1:
$$\frac{vg}{+}$$
$$\downarrow$$

F_2: $\quad 1\ \dfrac{+}{+} : 2\ \dfrac{vg}{+} : 1\ \dfrac{vg}{vg}$ (vg recessive to +).

(ii) Second and fourth pair show sex-linked inheritance for miniature (mn). As neither parent is miniature, and no vestigial miniature type exists, vestigial must be epistatic to miniature.

$$\frac{vg\ mn}{vg\ mn} \times \frac{+\ +}{+\ M}$$

$$\downarrow$$

F_1: $\quad\qquad ♀\dfrac{vg\ mn}{+\ +}\quad ♂\dfrac{vg\ mn}{+\ M}\quad$ (vg and mn recessive to +)

$$\downarrow$$

F_2 (phenotypes only given):

♀ gamete	♂ gamete			
	vg mn	+ mn	vg M	+ M
vg mn	♀ vestigial	♀ miniature	♂ vestigial	♂ miniature
vg +	♀ vestigial	♀ wild type	♂ vestigial	♂ wild type
+ mn	♀ miniature	♀ miniature	♂ miniature	♂ miniature
+ +	♀ wild type	♀ wild type	♂ wild type	♂ wild type

This gives F_2 ratios of:

> ♀ : 3 wild-type : 3 miniature-winged : 2 vestigial-winged
> ♂ : 3 wild-type : 3 miniature-winged : 2 vestigial-winged

which is close to those obtained for the fourth pair.

(iii) The third pair must be:

$$\frac{vg\ +}{vg\ mn} \times \frac{+\ +}{+\ M}$$

which give an F_1 as follows:

$$♀\frac{vg\ +}{+\ +}\ \text{and}\ \frac{vg\ mn}{+\ +}\ \text{i.e. all wild-type}$$

$$♂\frac{vg\ +}{+\ M}\ \text{and}\ \frac{vg\ mn}{+\ M}\ \text{i.e. 1 : 1 for wild-type and miniature.}$$

12. (i) Ignoring shades of colour, the scheme can be resolved as follows:

$$A \text{ or } B \times C \text{ or } D \quad \rightarrow \quad \text{coloured-flowered progeny}$$
$$A \times B \text{ or } C \times D \quad \rightarrow \quad \text{white-flowered progeny}$$
$$E \times A, B, C \text{ or } D \quad \rightarrow \quad \text{white-flowered progeny.}$$

It can be concluded that stocks A and B must be homozygous for a recessive mutant allele of the same gene. Similarly C and D, but that the two genes must be different, i.e.

$$\text{stock } A \text{ and } B \quad \frac{a+}{a+}$$

$$\text{stock } C \text{ and } D \quad \frac{+b}{+b}$$

$$\text{stock } E \text{ could be } \frac{a\,b}{a\,b}.$$

(ii) Now, considering differences leading to the production of red or mauve pigment, only $B \times D$ gives red, and so it can be proposed that these stocks are homozygous for a recessive mutant allele (c) of a third gene, whose wild-type allele is necessary for the conversion of red to mauve pigment and which is hypostatic to the a and b genes. The scheme now becomes:

$$A \; \frac{a++}{a++} \qquad B \; \frac{a+c}{a+c}$$

$$C \; \frac{+b+}{+b+} \qquad D \; \frac{+bc}{+bc}$$

$$E \text{ could be } \frac{a\,b\,c}{a\,b\,c} \left(\text{or } \frac{+c}{+c} \text{ or segregate for } c \text{ and } +c \right).$$

The possible biochemical basis might be:

$$\begin{array}{ccccccc}
\text{colourless} & \longrightarrow & \text{colourless} & \longrightarrow & \text{red} & \longrightarrow & \text{mauve.} \\
\text{precursor} & & \text{precursor} & & \text{precursor} & &
\end{array}$$

$$\begin{array}{ccc}
\text{blocked in } \dfrac{a}{a} & \text{blocked in } \dfrac{b}{b} & \text{blocked in } \dfrac{c}{c} \\
\text{individuals} & \text{individuals} & \text{individuals}
\end{array}$$

13. (i) Neither parent requires adenine, but $\frac{1}{4}$ of the progeny (58 out of 220) do. Two genes appear to be involved, and these must be unlinked. There are two possible explanations:

(a)

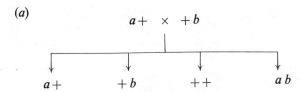

Only the doubly mutant strain required adenine (c.f. explanation to problem 8, and hairy versus hairless in problem 9).

A biochemical explanation would be:

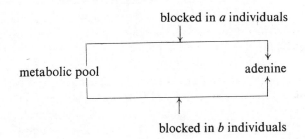

(b)

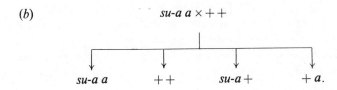

where mutation *a* leads to a requirement for adenine. The *su-a* mutation on its own has no effect, but combined with *a*, suppresses it. Such a situation would exist if *a* is a super-repressed (i^s) allele of a regulator gene, and *su-a* is an operator-constitutive (o^c) mutation (see page 161), but there are plenty of alternative explanations.

(c) Since conidiospore colour and the adenine requirement are not inherited independently, linkage must be involved. For scheme (a), the cross would be:

$$a\ y+ \times + \ +b \ (\text{or } ya \ + \times ++ \ b).$$

to give the following types of progeny.

	y and b relationship	conidiospore colour	requirement for adenine?
a y+	Parental	yellow	No
+ +b	Parental	green	No
+ y+	Parental	yellow	No
a +b	Parental	green	Yes
a yb	Recombinant	yellow	Yes
+ ++	Recombinant	green	No
+ yb	Recombinant	yellow	No
a ++	Recombinant	green	No

Only two of these classes can be identified ($a +b$ and $a yb$) but these are respectively parental and recombinant and can be used to calculate a recombination fraction i.e. $(2 \times 100)/(56 + 2) = 3.4\%$.

For scheme (b), the cross would be $su\text{-}a +a \times + y+$ and the recombination fraction for y and a can be calculated in the same way.

14. (i) The F_1 progeny have the wild-type eye colour, and so the mutant genes in the two stocks must be different and complementary.

(ii) The F_2 progeny show differences in eye colour segregation correlated with sex, and so sex linkage is involved.

(iii) Since among the F_2 progeny there are individuals with bright-red eyes and with brown eyes, genetic information for these characters must be present in the original parents.

(iv) The F_2 females show a segregation of 9 : 3 : 3 : 1 for the classes of eye colour involved, which suggests that two unlinked genes are involved.

(v) The ratio of F_2 males is not so obvious, but note that there are half the number of wild-type males, as wild-type females, and also that there are three times as many wild-type males, as either bright-red eyed or brown-eyed males. The ratio of white to coloured among males is 358 : 316, which is not very close to either a 1 : 1 (337 : 337) or a 9 : 7 (379 : 295). However if half the male progeny were white because of a sex-linked gene, the remainder would show a segregation of 188 wild-type : 62 red-eyed : 66 brown-eyed : 21 white-eyed, i.e. a 9 : 3 : 3 : 1 ratio.

It can be postulated that one stock is white-eyed because it is homozygous for alleles in two unlinked genes, which would independently cause either bright-red eyes or brown eyes (c.f. explanation given for problem 6). The other stock is white-eyed because of a single sex linked gene, which prevents any pigment formation, and will

$$\text{♀}\ \frac{brd\ bwn\ +}{brd\ bwn\ +} \times \text{♂}\ \frac{+\ +\ w}{+\ +\ M}$$

$$\downarrow$$

F₁:

$$\text{♀}\ \frac{brd\ bwn\ +}{+\ +\ w}$$

$$\text{♂}\ \frac{brd\ bwn\ +}{+\ +\ M}\ \left.\begin{array}{c}\\\\\end{array}\right\}\ \begin{array}{l}\text{phenotype}\\\text{dark-brown}\\\text{(wild-type)}\end{array}$$

$$\downarrow$$

F₂ (phenotypes only given):

gametes from ♀	brd bwn +	brd + +	+ bwn +	+ + +	brd bwn M	brd + M	+ bwn M	+ + M
brd bwn +	♀ W	♀ BR	♀ B	♀ DR	♂ W	♂ BR	♂ B	♂ DR
brd + +	♀ BR	♀ BR	♀ DR	♀ DR	♂ BR	♂ BR	♂ DR	♂ DR
+ bwn +	♀ B	♀ DR	♀ B	♀ DR	♂ B	♂ DR	♂ B	♂ DR
+ + +	♀ DR	♀ DR	♀ DR	♀ DR	♂ DR	♂ DR	♂ DR	♂ DR
brd bwn w	♀ W	♀ BR	♀ B	♀ DR	♂ W	♂ W	♂ W	♂ W
brd + w	♀ BR	♀ BR	♀ DR	♀ DR	♂ W	♂ W	♂ W	♂ W
+ bwn w	♀ B	♀ DR	♀ B	♀ DR	♂ W	♂ W	♂ W	♂ W
+ + w	♀ DR	♀ DR	♀ DR	♀ DR	♂, W	♂ W	♂ W	♂ W

W = white, B = brown, BR = bright red and DR = dark red.

therefore be epistatic to the mutant genes in the other stock. If the wild-type alleles are designated +, and the mutant alleles *brd, bwn, w* respectively, the scheme is as on page 201.

As sex linkage is involved the reciprocal cross will give different results:

$$\female \frac{+\ +\ w}{+\ +\ w} \times \male \frac{brd\ bwn\ +}{brd\ bwn\ M}$$

$$\downarrow$$

F_1:

$$\female \frac{+\quad +\quad w}{brd\ bwn\ +} \text{ dark-red (wild-type)}$$

$$\male \frac{+\quad +\quad w}{brd\ bwn\ M} \text{ white-eyed}$$

F_2: Full details will not be given, but the segregation obtained will be

same for each sex *i.e.* 17 white $\left(16 \text{ sex-linked} + 1 \dfrac{brd\ bwn}{brd\ bwn}\right)$: 9 dark-red :

3 bright-red : 3 brown.

The possible biochemical basis of these findings is as follows:

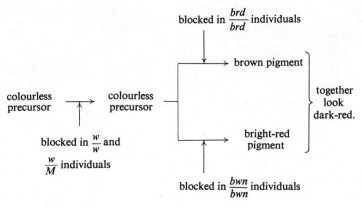

blocked in $\dfrac{brd}{brd}$ individuals

→ brown pigment

colourless precursor → colourless precursor

blocked in $\dfrac{w}{w}$ and $\dfrac{w}{M}$ individuals

bright-red pigment

together look dark-red.

blocked in $\dfrac{bwn}{bwn}$ individuals

References and recommended further reading

There are several more advanced textbooks covering the field described in this book. To be recommended are:

WILLIAM HAYES. *The Genetics of Bacteria and their Viruses*. Published by Blackwell Scientific Publications, Oxford and Edinburgh. 2nd edition 1968. Not only contains an authoritative account of bacterial and bacteriophage genetics, but the introductory chapters give an excellent treatment of more general genetic topics. It also contains a very comprehensive list of references to original papers.

JOHN MCLEISH AND BRIAN SNOAD. *Looking at Chromosomes*. Published by Macmillan and Co. Ltd, London, 1958. This inexpensive book contains an extremely good photographic account of both mitosis and meiosis.

M. W. STRICKBERGER. *Genetics*. Published by the Macmillan Co., New York, 1968. A good general genetics textbook.

JAMES D. WATSON. *Molecular Biology of the Gene*. Published by W. A. Benjamin, Inc., New York and Amsterdam, 2nd edition 1970. An up-to-date account of molecular genetics.

Complementary to this textbook, and covering the whole field of cell biology, the following two textbooks are to be recommended:

E. J. AMBROSE AND DOROTHY M. EASTY. *Cell Biology*. Published by Thomas Nelson and Sons Ltd, London, 1970.

ARIEL G. LOEWY AND PHILIP SIEKEVITZ. *Cell Structure and Function*. Published by Holt, Rinehart and Winston, London, New York, Sidney and Toronto, 2nd edition, 1970.

Below is a list of references to work described in this book. Reference is sometimes made to a later review paper rather than the first paper published on a topic. The numbers in brackets refer to the page in this book on which the work is cited.

AVERY, O. T., MACLEOD, C. M. AND MCCARTY, M. (1944). Studies on the chemical nature of the substance inducing transformation of pneumococcal types. I Induction of transformation by deoxyribonucleic acid fraction isolated from pneumococcal type III. *J. exp. Med.* **79**, 137. (64)

BATESON, W., SAUNDERS, E. R. AND PUNNETT, R. C. (1905). Experimental studies in the physiology of heredity. *Rep. Evol. Comm. R. Soc.* **2**, 1 and 80. (21)

BEADLE, G. W. AND TATUM, E. L. (1941). Genetic control of biochemical reactions in *Neurospora*. *Proc. nat. Acad. Sci., Wash.* **27**, 499. (4)

BENZER, S. (1961). On the topography of genetic fine structure. *Proc. nat. Acad. Sci., Wash.* **47**, 403. (106)

BONNER, J. (1965). *The Molecular Biology of Development.* Oxford: Clarendon Press. (181)

BRENNER, S. (1966). Colinearity and the genetic code. *Proc. Roy. Soc. Lond. B.* **164**, 170. (133)

BRENNER, S., STRETTON, A. O. W. AND KAPLAN, S. (1965). Genetic code: the 'nonsense' triplets for chain termination and their suppression. *Nature, Lond.* **206**, 994. (148)

BRIGGS, R. AND KING, T. J. (1959). Nucleocytoplasmic interactions in eggs and embryos. p. 537 in *The Cell*, vol. 1, ed. J. Brachet and A. E. Mirsky, New York: Academic Press. (177)

CAIRNS, J. (1963). The bacterial chromosome and its manner of replicating as seen by autoradiography. *J. molec. Biol.* **6**, 208. (90)

CHARGAFF, E. (1955). Isolation and composition of the deoxypentose nucleic acids and of the corresponding nucleoproteins. p. 307 in *The Nucleic Acids*, vol. I, ed. E. Chargaff and J. N. Davidson, New York: Academic Press. (69)

COEN, D., DEUTSH, J., NETTER, P., PETROCHILO, E. AND SLONIMSKI, P. P. (1970). Mitochondrial Genetics. I Methodology and Phenomenology. In *Control of Organelle Development*, S.E.B. Symposium 24, ed. P. L. Miller, London: Cambridge University Press.(120)

CRICK, F. H. C. (1958). On protein synthesis. *Symp. Soc. exp. Biol.* **12**, 138 (128)

CRICK, F. H. C., BARNETT, L., BRENNER, S. AND WATTS-TOBIN, R. J. (1961). General nature of the genetic code for proteins. *Nature, Lond.* **192**, 1227. (137)

DELBRÜCK, M. AND BAILEY, W. T., JNR. (1946). Induced mutations in bacterial viruses. *Cold Spr. Harb. Symp. quant. Biol.* **11**, 33. (99)

FRAENKEL-CONRAT, H. and SINGER, B. (1957). Virus reconstitution: combination of protein and nucleic acid from different strains. *Biochim. biophys. Acta* **24**, 541. (67)

GILBERT, W. AND MÜLLER-HILL, B. (1966). Isolation of the *lac* repressor. *Proc. nat. Acad. Sci., Wash.* **56**, 1891. (162)

GRIFFITH, F. (1928). Significance of pneumococcal types. *J. Hyg., Camb.* **27**, 113. (64)

GURDON, J. B. (1966). The cytoplasmic control of gene activity. *Endeavour*, **25**, 95. (178)

HADORN, E. (1968). Transdetermination in cells. *Scientific American* **219**, 110. (179)

HÄMMERLING, J. for a review of Hämmerling's work see Brachet, J. L. A. (1965). *Acetabularia. Endeavour* **24**, 155. (124)

HARRIS, H. (1968). *Nucleus and Cytoplasm.* Oxford: Clarendon Press. (182)

HERSHEY, A. D. (1946). Spontaneous mutations in bacterial viruses. *Cold Spr. Harb. Symp. quant. Biol.* **11**, 67. (99)

HERSHEY, A. D. AND CHASE, M. (1952). Independent functions of viral protein and nucleic acid in growth of bacteriophage. *J. gen. Physiol.* **36**, 39. (66)

HERSHEY, A. D., DIXON, J. AND CHASE, M. (1953). Nucleic acid economy in bacteria infected with bacteriophage T2: I. Purine and pyrimidine composition. *J. gen. Physiol.* **36**, 777. (126)

HOLLEY, R. W., APGAR, J., EVERETT, G. A., MADISON, J. T., MARQUISER, M., MERRILL, S. H., PENSWICK, J. R. AND ZAMIR, A. (1965). Structure of a ribonucleic acid. *Science N. Y.* **147**, 1462. (128)

INGRAM, V. W. (1957). Gene mutations in human haemoglobin: the chemical difference between normal and sickle cell haemoglobin. *Nature, Lond.* **180**, 326. (132)

JACOB, F. AND MONOD, J. (1961). Genetic regulatory mechanisms in the synthesis of proteins. *J. molec. Biol.* **3**, 318. (156)

JACOB, F. AND WOLLMAN, E. L. (1961). *Sexuality and the Genetics of Bacteria.* New York: Academic Press. (87)

KHORANA, H. G. See Nishimura, S., Jones, D. S. and Khorana, H. G.

LEDERBERG, J. AND TATUM, E. L. (1946). Novel genotypes in mixed cultures of biochemical mutants of bacteria. *Cold Spr. Harb. Symp. quant. Biol.* **11**, 1113. (85)

MENDEL, G. (1865). Translations of Mendel's two papers on experiments in plant hybridisation, originally published in *Verhandlungen naturforschender Verein* in Brünn, Abhandlungen, IV, will be found beginning on p. 419 in Sinnott, E. W., Dunn, L. C. and Dobzhansky, T. *Principles of Genetics*, New York, Toronto and London: McGraw-Hill Book Co. Inc., 5th edition 1958. (11)

MESELSON, M. AND STAHL, F. W. (1958). The replication of DNA in *Escherichia coli. Proc. nat. Acad. Sci., Wash.* **44**, 671. (70)

MÜLLER, H. J. (1928). The production of mutations by X-rays. *Proc. nat. Acad. Sci., Wash.* **14**, 714. (74)

NILSSON, L. R. (1967). On the ontogeny of alkaline phosphatases in *Drosophila melanogaster. Drosophila* Information Service, **42**, 60. (172)

NIRENBERG, N. AND LEDER, P. (1964). RNA codewords and protein synthesis. The effect of trinucleotides upon the binding of sRNA to ribosomes. *Science N. Y.* **145**, 1399. (145)

NIRENBERG, M. W. AND MATTHEI, J. H. (1961). The dependence of cell-free protein synthesis in *Escherichia coli* upon naturally occurring or synthetic polyribonucleotides. *Proc. nat. Acad. Sci., Wash.* **47**, 1588. (144)

NISHIMURA, S., JONES, D. S. AND KHORANA, H. G. (1965). Studies on polynucleotides XLVIII: The *in vitro* synthesis of a co-polypeptide containing two amino acids in alternating sequence dependent upon a DNA-like polymer containing two nucleotides in alternating sequence. *J. molec. Biol.* **13**, 302. (144)

PAUL, J., GILMOUR, R. S., THOMOU, H., THRELFALL, G. AND KOHL, D. (1970). Organ-specific gene masking in mammalian chromosomes. *Proc. Roy. Soc. Lond. B.* **176**, 277. (181)

SLONIMSKI, P. (1970). See Coen, D., Deutsh, J., Netter, P., Petrochilo, E. and Slonimski, P. P.

SONNEBORN, T. M. (1970). Gene action in development. *Proc. Roy. Soc. Lond. B.* **176**, 347. (187)

STERN, C. (1931). Zytologisch-genetische Untersuchungen als Beweisse für die Morgansche Theorie des Faktorenaustauchs. *Biol. Zbl.* **51**, 547. (39)

STEWARD, F. C. (1968). *Growth and Organisation in Plants.* Reading, Massachusetts: Addison-Wesley. (177)

TIEDEMANN, H. (1968). Factors determining embryonic differentiation. *J. Cell. Physiol.* **72** *Sup. 1*, 129. (187)

VOLKIN, E. AND ASTRACHAN, L. (1957). RNA metabolism in T2-infected *Escherichia coli.* p. 686. In *The Chemical Basis of Heredity.* Ed. W. D. McElroy and B. Glass. Baltimore: Johns Hopkins Press. (126)

WATSON, J. D. AND CRICK, F. H. C. (1953). The structure of DNA. *Cold Spr. Harb. Symp. quant. Biol.* **18**, 123. (69)

WILKINS, M. F. H., STOKES, A. R. AND WILSON, H. R. (1953). Molecular structure of deoxypentose nucleic acids. *Nature, Lond.* **171,** 738. (69)

YANOFSKY, C. (1963). Amino acid replacements associated with mutation and recombination in a gene and their relationship to *in vitro* coding data. *Cold Spr. Harb. Symp. quant. Biol.* **28**, 581. (108)

YANOFSKY, C., CARLTON, B. C., GUEST, J. R., HELINSKI, D. R. AND HENNING, U. (1964). On the colinearity of gene structure and protein structure. *Proc. nat. Acad. Sci., Wash.* **51**, 266. (133)

ZINDER, N. D. AND LEDERBERG, J. (1952). Genetic exchange in *Salmonella. J. Bacteriol.* **64**, 679. (95)

Index of definitions and glossary

Wherever possible, the technical genetic terms used in this book have been defined. The following is an index of these definitions, which because they appear in context, it is hoped will prove more satisfactory than a formal glossary.

Index

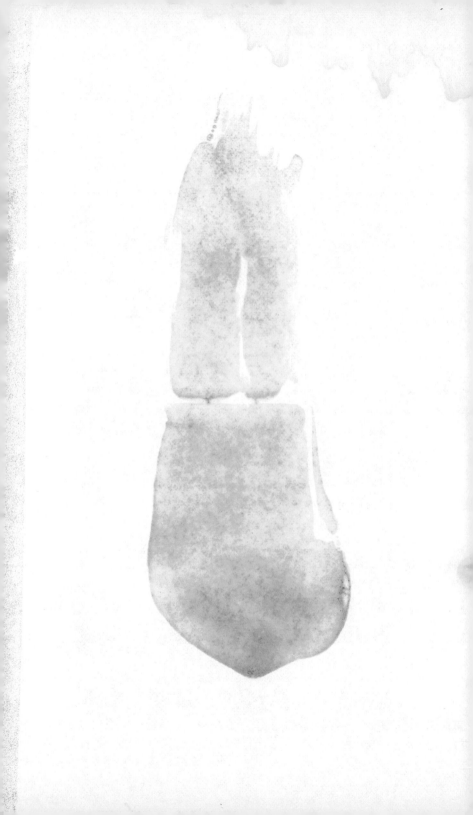

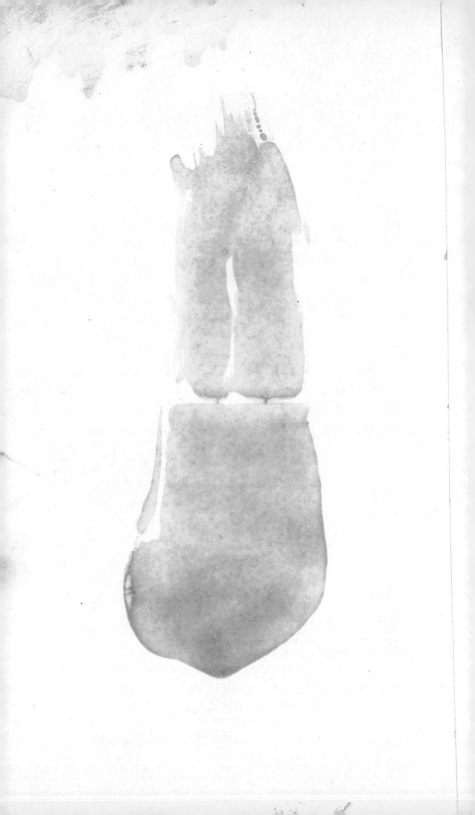